UK Military R&D

Report of a Working Party
Council for Science and Society

Oxford New York Tokyo
OXFORD UNIVERSITY PRESS
1986

Oxford University Press, Walton Street, Oxford OX2 6DP

Oxford New York Toronto
Delhi Bombay Calcutta Madras Karachi
Kuala Lumpur Singapore Hong Kong Tokyo
Nairobi Dar es Salaam Cape Town
Melbourne Auckland

and associated companies in
Beirut Berlin Ibadan Nicosia

Oxford is a trade mark of Oxford University Press

Published in the United States
by Oxford University Press, New York

British Library Cataloguing in Publication Data
UK military R&D: report of a working party
1. Military research — Great Britain.
I. Council for Science and Society
355'.07'0941   U395.G7
ISBN 0-19-859930-7

Library of Congress Cataloging in Publication Data
UK military R&D: report of a working party, Council
for Science and Society.
Bibliography: p.
1. Military research — Great Britain. I. Council for
Science and Society. II. Title. III. Title: UK military
R&D.
U395.G7U5   1986   355'.07'0941   85-29032
ISBN 0-19-859930-7 (pbk.)

Set by Express Litho Service (Oxford)
Printed in Great Britain by
Butler & Tanner Ltd, Frome and London

About the Council

The Council for Science and Society, a registered charity, was formed in 1973 with the object of 'promoting the study of, and research into, the social effects of science and technology, and of disseminating the results thereof to the public'.

The Council's primary task is to stimulate informed public discussion in the field of 'the social responsibility of the scientist'. It seeks to identify developments in science and technology whose social consequences lie just over the horizon, where no full-scale debate has yet begun. Experience shows that intensive analysis of the present (and probable future) 'state of the art', and of the foreseeable social consequences, can suggest a range of possible responses to those who will sooner or later have to take the necessary decisions. The Council carries out this task in a number of different ways, including the organization of conferences, seminars, and colloquia.

Major studies are conducted by *ad hoc* working parties, composed – like the Council itself – of experts in the respective fields together with lawyers, philosophers, and others who can bring a wide range of skills and experience to bear on the subject. The results of these studies are published in the form of reports such as this one. It is the Council's hope that these will help others to work out the most appropriate solutions to these problems in the course of responsible public debate, conducted at leisure on the best information available, rather than by the hurried, ill-informed, and ill-considered process which is apt to occur if the community does not become aware of a problem until it is too late.

The Council is grateful to the Trustees of the Esmée Fairbairn Trust, the Nuffield Foundation, and the Wates Foundation for financial support.

The Council welcomes all suggestions of further subjects for study.

COUNCIL FOR SCIENCE AND SOCIETY,
3/4 St Andrews Hill,
London EC4V 5BY
Tel: 01-236-6723

Council Members

Membership of the Working Party

Professor John Ziman, FRS Chairman
Chairman of the Council for Science and Society;
Visiting Professor in the Department of Social and Economic Studies,
Imperial College, London
Sir Hugh Beach
Retired Chief Royal Engineer/Former Master General of the Ordnance
Dr Mike Cooley
Greater London Enterprise Board
Dr Martin Edmonds
Department of Politics, University of Lancaster
Dr Philip Gummett Rapporteur
Department of Science and Technology Policy, University of Manchester
Professor W. Gutteridge
University of Aston
Dr Keith Hartley
Director, Institute for Research in the Social Sciences, University of York
Ms Scilla McLean
Director, Oxford Research Group
Professor Peter Nailor
Royal Naval College
Dr Julian Perry-Robinson
University of Sussex

Secretaries

Barbara Farah
Brigitte Gohdes

Foreword

John Ziman, FRS, Chairman of the Council for Science and Society and Visiting Professor in the Department of Social and Economic Studies, Imperial College, London

More than a quarter of the total scientific effort of the United Kingdom goes into the making of weapons of war. Thousands of scientists and engineers work on the conception, design, and development of fighters and frigates, tanks and communication systems, guided missiles, nuclear warheads, and many other items of military equipment. The security and the prosperity of the nation are vitally affected by this immense activity.

There is discussion in many circles in Britain on the scope and scale of military research and development – military 'R&D'. How does it fit into the strategies of NATO? What is its effect on other aspects of British foreign policy? What part does it play in the nuclear arms race and in the international arms trade? What commercial return does it yield, directly or indirectly, to British industry? Is it now one of the most important sources of all technological innovation? How does it relate to other R&D activities, in industry, government, or academia? What sort of a career is it, and how does it affect the supply of skilled personnel for advanced science and technology? These are only some of the many issues in which military R&D plays a major part – issues which have undoubtedly grown in urgency and importance over the last few years.

Like war itself, military R&D has thus become too serious a matter to be left to the generals, the chief scientific officers, and the undersecretaries of state. It is also highly controversial. Unfortunately, it is usually regarded as a completely esoteric activity, shrouded from the general view as much by its own technical complexity as by official confidentiality. Strong opinions about its place in the polity, the economy, or the scientific enterprise have frequently been formed without a background of knowledge of how it is actually undertaken and what it can achieve in practice.

These are points that we have been turning over in the Council for Science and Society for some years, both as very good reasons for producing a report such as this and as equally good reasons why such a report would be very difficult to do satisfactorily. We also foresaw – correctly, as it happens – that we were unlikely to get a specific grant to cover the cost of a working party on this subject, even though it was widely recognized to be one of the most important effects of science and technology on contemporary society. Nevertheless, in 1982 the Council decided to go ahead with such a study as part of our general

ix

programme, which is most generously supported by the Nuffield Foundation, the Esmée Fairbairn Trust, and (since January 1985) the Wates Foundation.

The decision was fully justified by the response we got from the people we approached to join the Working Party. Despite their wide range of experience, interests, and opinions, they did indeed work together as a party, complementing and courteously criticizing each others' highly expert contributions. We were all of us particularly grateful to Philip Gummett, who put an immense effort into weaving this diversity of views into a report that expresses them fairly and yet reads easily. Philip Gummett himself gratefully acknowledges the advice of a number of present and former officials in the Ministry of Defence, and the support of the Nuffield Foundation for providing a small grant during 1983/84 which accelerated his research on this subject. The Working Party would particularly like to thank Sir Clifford Cornford, former Chief of Defence Procurement in MoD, and Group Captain R. L. Easterbrook for very instructive discussions in the early stages of the preparation of this report. It would also like to thank Sir Robert Telford, President of The Marconi Co. Ltd., Dr Judith Reppy and Mr Darren Chadwick for helpful consultations. Naturally, none of those named bears any responsibility for the final report. Special thanks are due to the Council's secretaries, Brigitte Gohdes (during the early stages of the report) and Barbara Farah, and to Oxford University Press for publishing this report.

Contents

1. Introduction — 1
   1.1. Introduction — 1
   1.2. The origins of a military R&D programme — 2

2. The UK commitment to military R&D — 7
   2.1. Overall expenditure — 7
   2.2. Distribution of effort — 9
   2.3. International comparison — 11

3. Organization of military R&D — 16
   3.1. Formulation of R&D programmes — 16
      3.1.1. Formal procedures — 20
   3.2. The performance of R&D — 22
      3.2.1. The defence research establishments — 23
      3.2.2. Industry's role — 25

4. Policy issues — 28
   4.1. Introduction: imperial heritage — 28
   4.2. Abandonment of the nuclear role? — 30
   4.3. Consequences of over-sized military R&D effort — 31
      4.3.1. Budgetary problems — 31
         More selectivity — 32
         Less autonomy — 33
      4.3.2. Making better use of industry — 34
         More competition — 34
         Other difficulties — 35
      4.3.3. International collaboration and standardization — 36
         Renewed impetus for collaboration — 37
   4.4. Economic effects of military R&D — 40
      4.4.1. Spin-off — 40
      4.4.2. Structural impact of military R&D — 42
      4.4.3. UK studies — 45
   4.5. The personnel factor — 47
   4.6. The arms trade — 48
   4.7. The ethical dimension — 49

5. Conclusions and recommendations — 53
   5.1. Size and efficiency of military R&D effort — 53

5.2. Relation to national science and technology policy   56
5.3. The arms race and arms control   57
5.4. Public accountability and secrecy   59

References   62

1 Introduction

1.1 INTRODUCTION

Military research and development (R&D) in Britain is a large and expensive activity. Taking together the R&D activities performed by or for the Ministry of Defence (MoD), and the privately funded ventures of industrial corporations, Britain ranks high in the international league table of military R&D effort. Half of British government R&D spending goes into the military sector, and Britain, like the USA, spends a markedly greater percentage of its gross domestic product in this sector than any other major R&D spenders of the Western world, including France, Germany, and Japan.

It is inevitable that spending on such a scale should attract public and political interest. Military preparations must always be of concern for any peace-minded nation. Many people, including many among the scientific community, who fully accept the need for defence in an insecure world, are nevertheless very unhappy about the immense scale of the resources devoted to this purpose.

British public interest in defence matters happens to be high at present. The drama of the Falklands War is still fresh in the public mind, drawing attention to the issues which arise from the combination of downward financial pressure on the defence budget, and the upwardly moving costs of military equipment. This combination of competing forces has been a perennial feature of post-war British defence policy at least since the mid-1950s, but its effects have become particularly acute today, for reasons which include growing interest in the possible military use of expensive new advanced technologies and the uncertainties raised for policy-makers by sharper debate over the future of British nuclear weapons and US nuclear weapons in Europe. In addition, there is the question of British involvement in the American Strategic Defense Initiative.

Yet the R&D component of defence policy has tended not to feature strongly in discussions either of British defence policy or of British science and technology policy. Our purpose in this report is to open up debate on this important subject, and to argue that publicly unexamined policies have evolved which call for critical discussion and consideration of possible changes of course. In a brief report such as this we can only take up very general issues, indicating the dangers and deficiencies of certain current practices and suggesting directions for new initiatives. But many of us are now convinced that this is a subject requiring urgent attention, and would strongly recommend that more detailed proposals for change be worked out and put into practice without further delay.

1

To put these issues into perspective, we first illustrate how a military R&D programme arises, examine where military R&D figures in national science and technology policy and in the economy more generally, and detail how it is organized. We then discuss the key issues in relation to present problems and future prospects. These issues include the following.

(*a*) The effects of Britain's imperial heritage upon military R&D.

(*b*) The possible effects of any abandonment of British nuclear weapons.

(*c*) The consequences of the scale of Britain's military R&D effort.

(*d*) The personnel and ethical dimensions.

(*e*) The arms trade.

(*f*) The problem of secrecy.

We conclude with some observations and recommendations about the size and efficiency of Britain's military R&D effort, the relation of that effort to national science and technology policy, the arms race and arms control, and public accountability.

1.2 THE ORIGINS OF A MILITARY R&D PROGRAMME

The public conception of a solitary mad genius, discovering the principle of a 'death ray' device with which the world may be conquered, is entirely fictitious. It is quite true that basic scientific discoveries – nuclear fission, semiconductors, electromagnetic waves, etc. – are essential ingredients of all advanced technologies, civil and military. But it is very rare indeed for a new weapon to be conceived as such in the R&D laboratory, and the numerous 'bright ideas' that scientists and others come up with all the time are ruthlessly sifted for their practicality before they get any further.

Where, then, do military R&D programmes originate? Who is involved in their conception and development? There is no standard answer to these questions; every case is different, involving a complex interplay of needs and potentials. But whose needs do we mean: those of the customers, i.e. the armed forces (and if so, which one?), or those of the suppliers – the defence industries and MoD R&D establishments? And how is the choice made between different potential solutions to the customer's problem? To illustrate some of the processes that occur in the life of a military R&D programme, we consider a fictional case based upon a current problem.

Forces assigned to NATO in general, and British ones in particular, operate in an environment the nature of which is determined by the military technology available to the superpowers. As a specific example, they need to be protected in the event of air attack within Europe. Such protection could be done with anti-aircraft defences (fighter aircraft, or anti-aircraft artillery or ground-launched missiles) or by knocking out the threat at source by counter-attacking the enemy air bases. Each option has its advantages and disadvantages. Some involve short-range, and some longer-range fire with different possible implications for crisis stability. Some involve many small-scale encounters and others relatively few attacks on enemy air bases: the cost-effectiveness of these alternatives must be assessed.

Moreover, a problem like this does not arise out of the blue. It has existed for a long time, as have (more or less adequate) solutions to it. So also have technological, organizational, and other proposals to improve upon these solutions. Nothing, however, remains fixed for long in military technology. Suppose the time were approaching when current equipment needed replacing, or that the scale or nature of the threat were perceived to have changed because, for example, of the appearance of a new generation of ground attack aircraft. This perception would probably arise first within that branch of the armed forces whose own task would, thereby, be made more complicated. Or suppose that a new means of combating existing or potential threats began to appear feasible. Perhaps a combination of technological developments, or one single new innovation, allowed a radically new solution to an old problem such as the avoidance of jamming of radar. In this case, the new potential might be first appreciated in a defence research establishment, such as the Royal Signals and Radar Establishment (RSRE), or in a defence equipment firm, such as Marconi or British Aerospace, or perhaps in a university department with experience of working for MoD.

From such sources, a new element enters the existing stream of ideas, problems, practices, and plans. It will float about there, and within the defence community, for a time, until somehow it gains enough currency to begin to be examined more systematically. Military officers, operating through machinery which will be described in the next section, will begin to formulate a 'Staff Target' (see Para. 3.1.1), which identifies broadly what the objective of a new equipment programme might be. They will do this in consultation with scientific advisers, industrial companies, and civilian officials within MoD, so as to get a sense of the technical, industrial, financial, and political feasibility of the proposal.

When they are satisfied, the proposal proceeds through the MoD's decision-making machinery, passing through a set sequence of decision points which will be explained in Para. 3.1. After each one it progresses to fuller, and more expensive, research, development, and testing. Many hundreds of scientists and engineers might thus become engaged on various aspects of the proposed system. The initial research might be done in a defence research establishment, up to the point of a working device demonstrating the soundness of the principle or design concept. This would then be transferred, for design and development work, to an industrial firm working under contract to MoD. The firm might also add an element of private venture capital to the project. The project is regularly scrutinized by senior civilian and military officials, including scientific and technological staff. Defence Ministers will also be involved in those projects which are likely to be particularly expensive, and the Cabinet where they are very costly or of high political sensitivity. The evolving equipment programme is routinely checked for consistency with defence policy generally, including the prevailing military strategy, resource constraints, political objectives (including arms control), alliance considerations, possible alternatives from external sources, technological feasibility, and the projected production schedule. The project may also be tailored to enhance its attractiveness to foreign purchasers. In principle, therefore, there is a process of continuous scrutiny, with a decision being taken at each 'milepost' to proceed (or not) to the next, and with subsequent progress understood to depend on broadly favourable answers to the questions that remained unresolved from earlier stages. In practice, unexpected technical, political, economic, or commercial problems are continually being encountered, disturbing the orderly timetable.

Our air defence problem might lead the Army to consider Staff Targets for better anti-aircraft weapons, and for ground-launched missiles for attacking enemy air bases. The RAF, however, might look for improvements in air-to-air missiles (for aerial combat), air-launched missiles (for attacking air bases), and surface-to-air missiles for point defence of air bases. So, the initial problem might generate studies of four possible solutions, in one of which two Services might find a common interest (Table 1.1). It might not be obvious initially which

Table 1.1.

Armed Service	Short-range solutions (anti-aircraft)	Longer-range solutions (anti-air base)
Army	Anti-aircraft artillery (gun or missile)	Ground-launched missile
RAF	Air-to-air missiles Anti-aircraft artillery (gun or missile)	Air-launched missile

of these offer the best solution, nor which combination would be best. Preliminary studies would be conducted by the Defence Staff (see Fig. 3.1), assisted by the Ministry's research establishments and industrial firms. The Defence Operational Analysis Establishment (the Ministry's research establishment specializing in systems analysis) might be asked to examine the options, and combinations of them. The questions that these people and organizations would be asking include the following.

What exactly is the threat?

What combinations of weapons, in what numbers, and deployed how, would meet it?

What technical and production problems would need to be overcome?

What technology already exists?

What would the worthwhile combinations cost?

What would be the implications in terms of military personnel, organization, and deployment?

What are realistic time-scales for progress?

What are the implications with respect to Britain's allies?

What are the prospects for collaborative work?

What about export sales potential?

But a military R&D project involves too many uncertainties to be programmed by rational optimization. It is an arena of competing groups, whose interests cluster round different sets of questions and responses to them. Firms with products under development that fitted any of the options might be expected to argue for them. They would also be concerned with the implications of possible options for the flow of work through their development and production facilities. Military officers might press for solutions that could be expected to enter operational service earlier rather than later, and must be expected to watch out for the implications of the options for the future of their own Service, and for its capacity to engage in combined operations with the other UK Services and Allies. In addition, they would have an eye to the compatibility of the equipment with evolving military doctrine, such as, in our example, the NATO doctrine of Follow-On Forces Attack – that is, striking at reinforcements deep behind the main battle area. Civilian officials would be concerned with budgetary implications and consistency with current political objectives, such as arms control, and commitment to international arms collaboration. Staff in government research establishments would press for the lines they were keen to explore. The MoD Chief Scientific Adviser would try to ensure that no relevant technical or operational argument had been overlooked and that thinking about technical options had been projected imaginatively and sufficiently far into the future. He might, for example, question whether or not certain types of weapons systems would have any place on the battlefield in the year 2010 – a date well within the operational lifetime of any major new weapons development of the late-1980s. Given the gestation period for new weapons systems, there might also be changes in the definition of the project arising from turnover among the MoD staff involved, especially the military staff who tend to have relatively short tours of duty in Whitehall.

Out of such a process as this would emerge a set of views as to which, or what mix, of the four options previously identified would be pursued to completion.

If, for example, the tactic of attack on enemy air bases were favoured, then gradually the characteristics of the appropriate missiles would be clarified (ground -or air-launched; range; payload; type of warhead), as would the numbers needed, the manufacturer and production schedule, and the operational implications.

Even when these apparently firm specifications have been agreed, there may be further deviation from the programme as a result of unexpected technical hitches or economic glitches. R&D activity does not stop with a production contract; it is woven into every stage of the weapons procurement process.

In the meantime, of course, long-range thinking about the threat and ideas about the next generation of equipment for dealing with it would already have begun. The agreed solution to the problem used in our example would by now have become part of the picture which subsequent proposals would have to take into account.

2 The UK commitment to military R&D

Some simple statistics will show the scale and nature of Britain's commitment to military R&D. No single measure can, of course, give a full picture of so complex an activity. Assessments of whether or not a country is doing too much, or too little, will vary with the measure chosen. But a variety of indicators combine to give an overall picture of Britain's commitment.

2.1 OVERALL EXPENDITURE

Figure 2.1 shows how *British spending on military R&D*, expressed in constant prices, has moved since the 1950s, and how it compares with that of its main economic competitors. Although its spending was inevitably much smaller than that of the USA, Britain was the only major European state to maintain a high and unbroken commitment to military R&D throughout the post-war decade. Germany and Italy (and Japan) were limited by treaty (as well as by inclination) from engaging in certain defence activities, while the French recovery (in military R&D terms) began only in the 1960s. R&D were high defence priorities of the Attlee Governments, with major research programmes in aircraft (such as the medium-range bombers), guided missiles, and, of course, the atomic bomb all being started, and moving from research to development in the late-1940s and early-1950s. Cancellations of missile and aircraft projects, the Defence Reviews of 1964–68, and, for the first time (at least in peace time), the purchase from foreign suppliers of major weapons systems (including Polaris missiles), are among the factors which are reflected in the fall from 1961. The upturn from 1970 is probably attributable in large part to the Tornado aircraft and the Chevaline improvement of the Polaris warhead. Despite the dip from 1980, Britain remains at the top of the second division among spenders on military R&D.

Figure 2.1 also shows the *percentage of the defence budget* spent on military R&D, again in international comparison. Britain is among those countries which have chosen to invest a very high proportion of their military spending on R&D. The figure also shows that the R&D share of the British defence budget rose when the defence budget rose and fell when it fell, so that it is a real movement and not a statistical artefact.

Figure 2.1 also suggests that spending on military R&D is highly concentrated in a small number of countries and this point is confirmed if data for a larger set of countries are examined.[1]

7

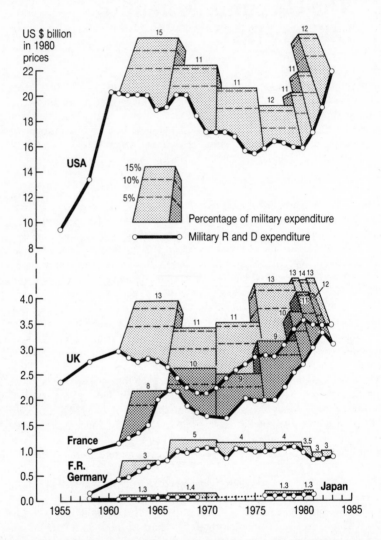

Fig. 2.1. Military R&D expenditure in selected countries in constant (1980) prices and as a percentage of total military expenditure. (Sources: Stockholm International Peace Research Institute (1972 and 1984). *SIPRI Yearbook 1972*, Table 6A.4 (for 1955 and 1958 data and for the UK 1961 data, all converted into 1980 prices). Almquist and Wiksell, Stockholm; *SIPRI Yearbook 1984*, Tables 6.2 and 6.4 Taylor and Francis, London.)

Finally, Fig. 2.2 shows the *percentage of gross domestic product* (GDP) spent on military R&D by Britain and its main economic rivals. These data confirm the impression that, by international standards, Britain is very heavily committed to military R&D.

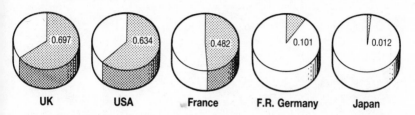

Fig. 2.2. Military R&D expenditure as a percentage of GDP, 1981. (Source: OECD data, cited in Cabinet Office (1984). *Annual review of Government funded R&D 1984*, Table 3.1. HMSO, London.)

2.2 DISTRIBUTION OF EFFORT

That commitment can, of course, be measured in terms of people as well as money. In 1984–85, some 4800 qualified scientists and engineers were directly employed by the Ministry of Defence on military R&D.[2] No comparable figures exist for industry. Figure 2.3 shows the distribution of *all MoD civilian R&D staff* (i.e. not just those with a degree or equivalent) by area of research, over time. One obvious feature of Fig. 2.3 is the reduction in the MoD's R&D staff that has occurred in recent years; the 1984–85 total of 25 900 (including all categories of staff, not simply scientists and engineers) is 32 per cent down on the 1975–76 total.

Figure 2.3 can usefully be compared with Fig. 2.4, which shows the *distribution of the military R&D budget* by the same areas of work as Fig. 2.3. The *staff* are heavily concentrated on 'other R&D'* and then on aircraft; by contrast far the biggest *financial* effort is in aircraft, with the rest, apart from ordnance and other army equipment, spread fairly evenly across the board. This is not surprising: aircraft development is intrinsically more expensive than ship and tank development.

We also see that the distribution of funds has remained fairly steady over the period of the data, though with some transfer between aircraft and the areas of ship construction and underwater warfare, coinciding with the rundown of effort

---

*Other R&D is a residual category. It includes work that is not attached to any of the main project areas, such as some of the more fundamental research. Its main component, however, is research and development of nuclear weapons.

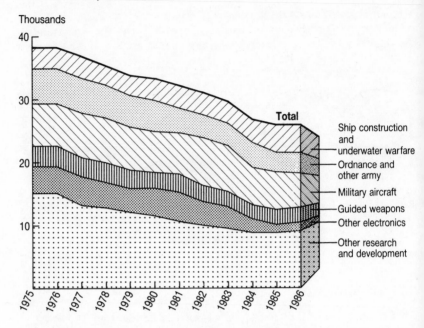

Fig. 2.3. Functional analysis of MoD civilian R&D staff, 1975–76 to 1985–86. (Sources: Ministry of Defence (1980 and 1985). *Defence in the 1980s: statement on the defence estimates 1980*, Cmnd. 7826-II, Vol. II, Table 5.2; *Statement on the defence estimates 1985*, Cmnd. 9430-II, Vol. II, Table 5.2. HMSO, London.)

on Tornado and the build-up of effort on submarine detection, reducing noise levels of Polaris submarines, and preparation for Trident.

Figure 2.5 shows *where the work is done*. In recent years a little over two-thirds has been spent outside the Ministry's own research establishments.

The overwhelming bulk of military R&D money has been spent on work done in private industry or public corporations, and this is mainly spent on development rather than research. In fact, about 80 per cent of the total military R&D budget goes on *development*, and only about 20 per cent on *research* (Fig. 2.8).

Moreover, only a few per cent of the intramural expenditure now goes on *basic* research. The Strathcona Committee in 1980, prompted by the MoD's Chief Scientist, Sir Ronald Mason, expressed its concern at the serious reduction of long-term research in the defence research establishments in the preceding years. As a consequence, there was a build-up of funds going to the universities from 1979 (with a hiccup in 1983–84), as shown in Fig. 2.5. Moreover, a co-operative grant scheme was instituted in 1985, under which

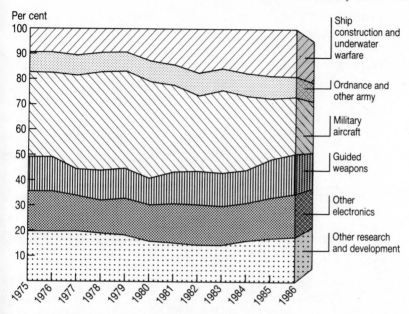

Fig. 2.4. Functional analysis of defence R&D budget, shown as percentage of total defence R&D spending, 1975–76 to 1985–86. (Sources: Ministry of Defence (1981 and 1985). *Statement on the defence estimates 1981*, Cmnd. 8212-II, Vol. 2, Table 2.4; *Statement on the defence estimates 1985*, Cmnd. 9430-II, Vol. 2, Table 2.3. HMSO, London.)

MoD provided £10 million for grants to universities, to be administered through the research councils.[3]

2.3 INTERNATIONAL COMPARISON

Finally, British spending on military R&D needs to be seen within the framework of national and governmental spending on R&D as a whole. Figure 2.6 shows what *percentage of total national R&D* spending goes to military R&D in a number of countries. Figure 2.7 shows what *percentage of government expenditure on R&D* (GERD) goes to military R&D in the same countries and Fig. 2.8 shows how British government spending on military R&D compares with the *rest of GERD* for 1981–82. These figures show a clear separation between the USA, UK, and France, on the one hand, and the Federal Republic of Germany, Italy, and Japan on the other, in terms of the commitment of national and governmental R&D funds to military R&D. They reinforce, therefore, our earlier remarks about the concentration of military R&D within a

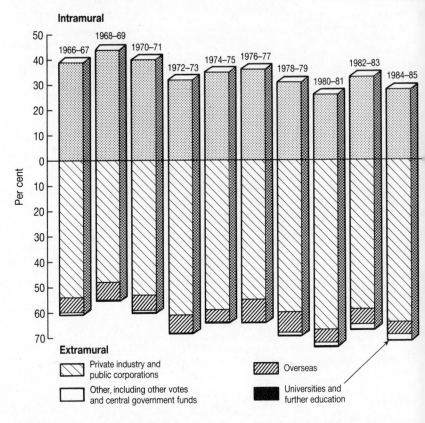

Fig. 2.5. Intra- and extramural distribution of UK defence R&D expenditure as percentage of total, 1966–67 to 1984–85. (Sources: Central Statistical Office (1973 and 1976). *Research and development expenditure*, Studies in Official Statistics series, No. 21 (1973, Table 6) and No. 27 (1976, Table 4b); *Economic Trends* (1979). *Economic Trends*, No. 309, Table 4; Ministry of Defence (1982 and 1985). *Statement on the defence estimates 1982*, Cmnd. 8529-II, Vol. 2, Table 3.2 and *Statement on the defence estimates 1985*, Cmnd. 9430-II, Vol. 2, Table 3.2. All the above are published by HMSO, London.)

few countries (including, of course, the USSR, on which figures have not been presented here for lack of systematic and comparable sources). We should also note that military R&D consumes a smaller share of the total national and governmental R&D resources of the USA, UK and France, than it did in the 1950s and early-1960s, though in all three cases it has risen again in recent years.

It is important to recognize that Fig. 2.6 reflects movements in three variables: military R&D spending, government spending on civil R&D, and

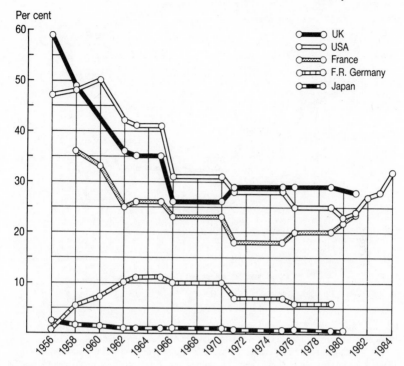

Fig. 2.6. Defence R&D as percentage of total national R&D spending. (Sources: Stockholm International Peace Research Institute (1972 and 1984). *SIPRI Yearbook 1972*, Table 6A.8. Almquist and Wiksell, Stockholm; *SIPRI Yearbook 1984*, Table 6.4. Taylor and Francis, London.)

industrial spending on R&D. For Britain, the balance between governmental civil and military spending is shown in Fig. 2.8. Figure 2.8 also illustrates the distinction drawn by MoD between its *research* and its *development* activities, and provides the basis for the argument that British military *research* spending is not excessive compared with the sums spent on research by civil departments. MoD is unique in Whitehall in its role as a purchaser of equipment, and it is from this role that its development budget flows. Other departments, in contrast, are usually engaged in sponsoring research which someone else will take into development and final use. Hence, unlike MoD, the civil departments do not spend heavily on development. The MoD therefore defends itself against the charge that it spends excessively on R&D by arguing that what should be compared is its *research* spending with the *research* spending of the civil departments.

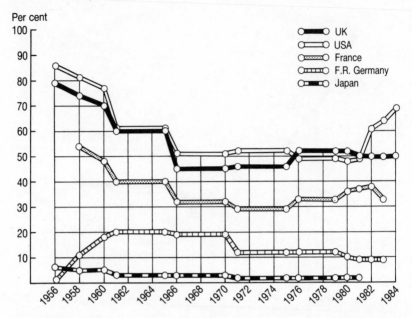

Fig. 2.7. Military R&D expenditure as percentage of government expenditure on R&D, 1956–84. (Sources: Stockholm International Peace Research Institute (1972 and 1984). *SIPRI Yearbook 1972*, Table 6A.7. Almquist and Wiksell, Stockholm; *SIPRI Yearbook 1984*, Table 6.3. Taylor and Francis, London; and (for UK 1984 data) Cabinet Office (1984). *Annual review of Government-funded R&D 1984*, p. 10. HMSO, London.)

Whatever the merits of this argument, it has been contended in scientific circles in recent years that spending on *civil* science has fallen too low. If this is true, it would make the percentage of military R&D in both national and governmental spending on R&D seem unduly large. It can also be suggested that industrial spending on R&D has fallen too low by the standards of Britain's industrial competitors. Organisation for Economic Co-operation and Development (OECD) figures[4] confirm this for manufacturing industry, though not for the service sector where Britain is a world leader (and, of course, the crucial question of what the optimal level is for each country cannot be resolved by such comparisons). Britain's service sector, which includes power and water supply, transport, communications, and business and commercial services, employs as many research scientists and engineers, in the same sector, as France and Germany combined. In the manufacturing sector, however, British industry provides a lower proportion of total national R&D funds than its major competitors. It emulates the USA (but not France, Germany, and Japan) in the extent to which R&D is concentrated in large firms and, in terms of R&D

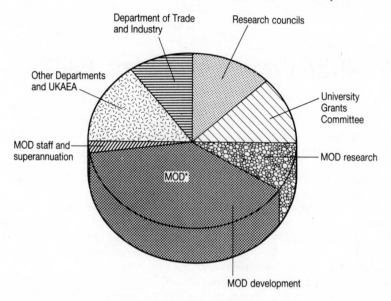

Fig. 2.8. Distribution of Government R&D expenditure, 1984–85. (Source: Adapted from Cabinet Office (1984). *Annual review of Government-funded R&D 1984*, Fig. 2.1. HMSO, London.)

spending per employee, it falls out of the major league and into the medium spending class, below Norway and near to Finland.

Even if we accepted that military R&D is excessively funded in Britain, it would not follow that if less money were spent on military R&D, more money would be spent on civil R&D. The practice of the British Government is to insist that government expenditure on R&D is not a totality, to be distributed centrally. Instead, each department of government is responsible for funding its own R&D activities, and a reduction in one department carries no implication for another. Rather than leading to more civil R&D, a reduced military R&D budget could lead to a different pattern of equipment purchasing, or to a smaller defence budget. Whether this decentralized system for managing governmentally-funded R&D is altogether appropriate is a different matter, and one to which we return in Para. 5.2.

3 Organization of military R&D

Military R&D in Britain is organized almost entirely from the Ministry of Defence. It is the MoD which sets the requirements for military equipment, and hence for R&D, and it is the MoD which oversees the implementation of the R&D programme, and subsequent testing and production. We therefore begin this chapter with a description of the relevant MoD procedures for setting and managing the military R&D programme. After this, we consider where the R&D is done, namely, in the defence research establishments and industry. First, however, we should note that the organization for military R&D in Britain has undergone many changes in the past few decades, as the central machinery for defence policy has itself changed. The most recent change took effect from January 1985. It resulted in the organization shown in Fig. 3.1. In Fig. 3.2 we give a brief indication of the roles played by the main elements of that organization which are relevant to UK military R&D.

3.1 FORMULATION OF R&D PROGRAMMES

As we saw in Para. 1.2, requirements for new military equipment, and hence also for R&D, arise from four main sources.

(*a*) A new assessment of the threat;

(*b*) The development of a new capability;

(*c*) A change in the concept of operations;

(*d*) Obsolescence of an existing item of equipment.

An argument based on any one of these positions may enter the decision-making system in a number of ways. It may come from scientists in one of the defence research establishments, or from an industrial contractor; it may come from a firm which has had no previous dealings with the Ministry of Defence, or from a university scientist; or it may come from one of the Services, or from the Defence Scientific Advisory Council. Indeed, to refer to 'the source' of a new project may be to commit a category mistake. It implies the injection of a new factor into an otherwise static picture. In reality, however, the Ministry of Defence runs on a complex set of interconnections between its parts. We have already described some of these interconnections as they arise at the highest levels, and it is not hard to imagine similar interconnections right down the line. Thus, there is constant interplay between the Services, the civilian secretariat and scientific staffs, and the staff of the Procurement Executive.

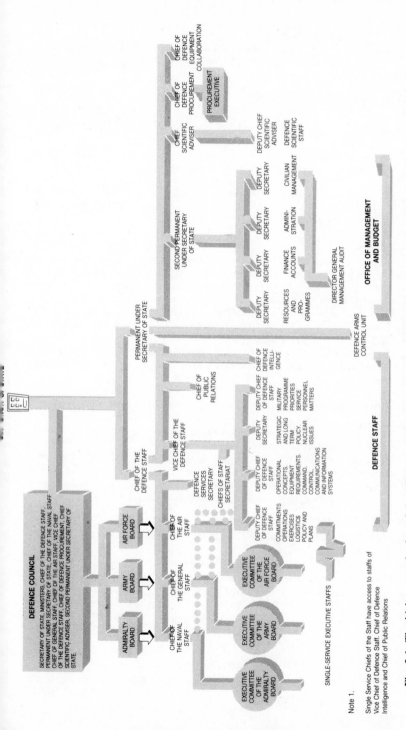

Fig. 3.1. The higher organization of the Ministry of Defence. (Source: Slightly adapted from Ministry of Defence (1984). *The central organisation for defence*, Cmnd. 9315. HMSO, London.)

DEFENCE COUNCIL
SECRETARY OF STATE, MINISTERS, CHIEF OF THE DEFENCE STAFF, PERMANENT UNDER SECRETARY OF STATE, CHIEF OF THE NAVAL STAFF, CHIEF OF GENERAL STAFF, CHIEF OF THE AIR STAFF, VICE CHIEF OF THE DEFENCE STAFF, CHIEF OF DEFENCE PROCUREMENT, CHIEF SCIENTIFIC ADVISER, SECOND PERMANENT UNDER SECRETARY OF STATE

ADMIRALTY BOARD

ARMY BOARD

AIR FORCE BOARD

CHIEF OF THE NAVAL STAFF

CHIEF OF THE GENERAL STAFF

CHIEF OF THE AIR STAFF

SINGLE-SERVICE EXECUTIVE STAFFS

EXECUTIVE COMMITTEE OF THE ADMIRALTY BOARD

EXECUTIVE COMMITTEE OF THE ARMY BOARD

EXECUTIVE COMMITTEE OF THE AIR FORCE BOARD

CHIEF OF THE DEFENCE STAFF

PERMANENT UNDER SECRETARY OF STATE

VICE CHIEF OF THE DEFENCE STAFF

DEFENCE SERVICES SECRETARY

CHIEFS OF STAFF SECRETARIAT

CHIEF OF PUBLIC RELATIONS

CHIEF OF DEFENCE INTELLIGENCE

DEPUTY CHIEF OF DEFENCE STAFF
OPERATIONAL REQUIREMENTS
COMMAND, CONTROL, COMMUNICATIONS AND INFORMATION SYSTEMS

DEPUTY CHIEF OF DEFENCE STAFF
OPERATIONAL CONCEPTS EQUIPMENT

DEPUTY CHIEF OF DEFENCE STAFF
COMMITMENTS OPERATIONS EXERCISES LOGISTICS POLICY AND PLANS

DEPUTY CHIEF OF DEFENCE STAFF
STRATEGIC AND LONG TERM POLICY, NUCLEAR ISSUES

DEPUTY SECRETARY
MILITARY PROGRAMME PRIORITIES SERVICE PERSONNEL MATTERS

DEFENCE ARMS CONTROL UNIT

DEFENCE STAFF

SECOND PERMANENT UNDER SECRETARY OF STATE

CHIEF SCIENTIFIC ADVISER

CHIEF OF DEFENCE PROCUREMENT

CHIEF OF DEFENCE EQUIPMENT COLLABORATION

PROCUREMENT EXECUTIVE

DEPUTY CHIEF SCIENTIFIC ADVISER

DEFENCE SCIENTIFIC STAFF

DEPUTY SECRETARY
RESOURCES AND PRO-GRAMMES

DEPUTY SECRETARY
FINANCE ACCOUNTS

DEPUTY SECRETARY
ADMINISTRATION

DEPUTY SECRETARY
CIVILIAN MANAGEMENT

DIRECTOR GENERAL MANAGEMENT AUDIT

OFFICE OF MANAGEMENT AND BUDGET

Note 1.

Single Service Chiefs of the Staff have access to staffs of Vice Chief of Defence Staff, Chief of Defence Intelligence and Chief of Public Relations

MAIN ELEMENTS OF MOD ORGANIZATION RELEVANT TO R&D

1. *Chief of the Defence Staff (CDS)*

One of the Secretary of State's two principal advisers (see Permanent Under Secretary of State, below), and the government's principal military adviser. Responsible for:

a) tendering military advice on strategy, forward policy, overall priorities in resource allocation, programmes, current commitments, and operations;
b) the planning, direction, and conduct of all national military operations;
c) directing the work of the Defence Staff (see below);
d) chairing the Chiefs of Staff Committee, on which sit the Chiefs of the three Armed Services.

1.1 *Defence Staff*

A new unified Defence Staff has, not uncontroversially, replaced the previous military staffs and the greater part of the previous Naval, General (i.e. Army), and Air Staffs. Its task is to assist the Chiefs of Staff to find the best solution to the problems of the day, whether of an operational, strategic planning, defence policy, or equipment nature. It contains military, civilian secretariat, and civilian scientific staffs. Although responsible to the CDS for all military aspects of its work, the Defence Staff is also responsible to the Permanent Under Secretary for the political and parliamentary aspects of its work and co-ordination with other government departments. Scientific staff within the Defence Staff are professionally accountable to the Chief Scientific Adviser (see below).

Within the Defence Staff, the principal grouping of relevance here is the *Systems* grouping, headed by a Service Deputy Chief of the Defence staff at 3-star level (Lt. General or equivalent: Deputy Secretary rank in the civil service). This group is responsible for the formulation of operational concepts, the determination and sponsorship of operational requirements, and setting the aims of the military research programme. It brings together the previously separate Service operational requirements organizations and those in the central staff (thus eliminating the need for the former Operational Requirements Committee). It is also responsible for command, control, and communications systems across all three services.

2. *Permanent Under Secretary of State (PUS)*

The permanent head of the Ministry and the second of the Secretary of State's two principal advisers. Responsible for:

a) the organization and efficiency of the Ministry including the management of all civilian staff and the co-ordination of its business;
b) the long-term financial planning and budgetary control of the defence programme, the associated allocation of resources, and the proper scrutiny of the requirements for all proposals with expenditure implications;
c) advice on the political and parliamentary aspects of the Ministry's work and relations with other Departments.

2.1 *Office of Management and Budget (OMB)*

One of the aims of the 1985 reorganization was to provide within the Ministry of

Defence much stronger central determination of priorities for expenditure and control of resource allocation. To this end, the PUS's responsibility for long-term financial planning and resource allocation and for the scrutiny and control of expenditure has been concentrated in an Office of Management and Budget.

Within OMB, the relevant element for our purposes is the one concerned with *Resources and Programmes*. This is responsible for co-ordinating the Ministry's annual long-term costing and for the Ministry's contribution to the Government's annual Public Expenditure Survey. It also scrutinizes major proposals for expenditure including new equipment programmes.

3. *Chief Scientific Adviser (CSA)*

Responsible to the Secretary of State through PUS for providing independent long-term thinking and scrutiny and for the central management of operational analysis (including oversight of the work of the *Defence Operational Analysis Establishment*, and encompassing also the work that previously was done by the three Service Chief Scientists, whose posts have lapsed). Assisted by a small central scientific staff. Further scientific staff who are deployed within the Defence Staff are professionally accountable to CSA. Is expected to provide a critical perspective upon proposals coming in from the established professionals.

4. *Chief of Defence Procurement (CDP)*

Responsible to the Secretary of State through PUS for the direction of the *Procurement Executive (PE)*, through which is spent about 40 per cent of the defence budget on the research, development, production, and purchase of military equipment. The PE is organized under four Controllers: *Controller of the Navy*, *Controller of the Air Force*, *Master General of the Ordnance* (that is, Controller of the Army), and *Controller of Establishments, Research, and Nuclear programmes (CERN)*. These are broadly responsible for matters to do, respectively, with Sea, Air, and Land Systems, and with the operations of the *research establishments* (see Fig. 3.4) and nuclear programmes. Prior to March 1985, CDP held responsibility for international equipment collaboration as well as for the general management of the defence procurement programme. In March 1985, however, the post of *Chief of Defence Equipment Collaboration* was created to concentrate full-time on the pursuit of collaboration.

5. *Principal Committees*

The principal committees which pull the military R&D activity together are:

a) *Financial Planning and Management Group:* chaired by PUS; contains the Chiefs of Staff and CDP. Advises the Secretary of State on main resource allocation decisions.

b) *Equipment Policy Committee:* chaired by CSA; contains senior officials from the OMB, the deputy CSA, members of the Defence Staff responsible for operational requirements and equipment, the four Controllers within the PE, and the Head of Defence Sales. Advises Ministers and the Chiefs of Staff on the equipment production and development programme, and the balance of equipment investment, so as to ensure that they are matched to operational requirements, resources, defence policy, industrial and sales considerations, and technical feasibility, and provide value for money. Supported by subcommittees dealing with individual areas of the equipment programme.

c) *Defence Research Committee:* chaired by CSA; contains members of the Defence Staff, the four Controllers within PE, and other scientific, operational requirements and defence policy staff. Monitors CERN's management of the research establishments, sets the broad objectives for the programme of long-term research (as distinct from development) and considers scientific and technical matters concerning defence policy generally.

d) *Defence Scientific Advisory Council:* a group of independent academic and industrial advisers plus some officials, chaired by a distinguished independent member. Offers independent advice to the Secretary of State for Defence in scientific and technological matters of concern to MoD. Limited in size to 30 members, including officials. Advises on the work at the research establishments in the light of relevant knowledge in the universities and industry, and scrutinizes the R&D programme without having any prior commitment to specific policies or projects. Usually holds about five or six formal meetings a year and visits two establishments. Subgroups hold separate meetings as necessary. Membership of subgroups and of the Council itself totals some 200 scientists, but only the chairman's name is published.

Fig. 3.2. Main elements of MoD organization relevant to R&D. (Sources: Ministry of Defence (1984). *The central organisation for defence*, Cmnd. 9315; House of Commons (1984). Third report from the Defence Committee Session 1983–84, *Ministry of Defence reorganisation*, HC 584. Both references published by HMSO, London.)

An officer fresh from Northern Ireland or the Falklands may be posted directly to a research establishment where his perceptions of existing equipment, tactical problems, changes in the threat, or the impending obsolescence of current equipment bump up against new scientific and technological enthusiasms and possibilities, are chastened by the doubts of experts about the feasibility of certain types of solution, and lead to continuous modification of existing programmes. Developments in rival equipment overseas, or in the strategic or economic environments, may be fed in from the field, from policy or intelligence quarters, or from the scientific community.

3.1.1 Formal procedures

Whatever the actual origin, for an idea to enter the equipment programme of the Ministry it must first be formally expressed as a *Staff Target* (see Para. 1.2), drawn up by the operational requirements staff of the Defence Staff. The Staff Target outlines, in broad terms, the function and desired performance of the item. It is prepared in consultation with the appropriate systems Controller (Land, Sea, or Air) in the Procurement Executive, the research establishments concerned, the Defence Science Staffs, and the relevant sections in industry.

To proceed further the project now requires formal approval. This approval will have to be sought again at every subsequent milepost. Depending on the expected cost of the project, it may require approval from the Equipment Policy

Committee (EPC), a subcommittee of the EPC, or simply the Deputy Chief of the Defence Staff (Systems). Projects expected to cost more than £50 million for development or £100 million for production will be in the first category, and cheaper ones will be in the second and third categories. Should the expected cost of a project change significantly during its lifetime, its approving authority will change also. There is, in addition, an increasing trend to refer to ministers all projects in the two most expensive categories.

Having been approved, the proposal then passes to the Procurement Executive where a *feasibility study* is carried out. (See Fig. 3.3 for the stages through which the proposal passes.) This is done using the research establishments and/or industry and is intended to allow for judgements to be made about the technical feasibility of the concept, to identify alternative options, and to prepare estimates of the work, time, and cost involved.

Once the feasibility study has been done, the Defence Staff prepares a *Staff Requirement*. This specifies in detail the required function and performance of the proposed equipment. Progress at this stage depends upon consistency with strategic concepts, harmonization wherever possible between the three Services and Britain's allies, compatibility with other planned and existing equipment, and, of course, an apparently sound financial and technological base.

The project then goes to the appropriate approving authority. If approved here, taking account of budget and manpower resources, industrial capabilities, and the scope for collaborative development or foreign purchase, then the proposal proceeds to the *Project Definition* stage. Except for very expensive projects, it is only at this stage that large sums of money begin to be committed, and hence further milestones are built into the procedures to try to ensure that

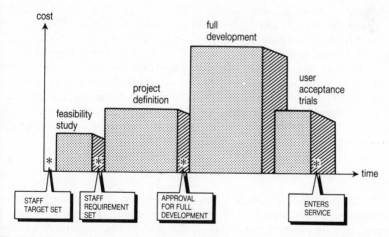

Fig. 3.3. Milestones in the progress of an equipment project: a schematic representation.

the rise in costs which now follows is appropriate to the activity involved and is not the product of mismanagement.

The Project Definition phase aims to investigate the degree of technical risk associated with the project and to arrive at a complete plan of action including a definitive assessment of the total R&D and production costs and time-scale. The possibilities for international collaboration are examined once more, and the personnel, training, and logistics implications considered. The results go for one or more further reviews by the appropriate approving authority to check whether anything has changed in the time (which could be several years) since they had endorsed the Staff Requirement. (For example, has the perceived threat changed; do the cost or other implications now make the project less attractive; or, as in much computer-based technology, has the state of the art changed dramatically during development?) The project may then be approved for *Full Development* including, if appropriate, prototype construction, followed by user trials and acceptance trials. Ministerial and Treasury approval will normally be needed at this stage if it has not already been sought at an earlier stage. A project manager is now appointed (indeed, may already have been appointed at the Feasibility Study stage for a very large project), and a full-blooded commitment is made to the project. A change of heart after this stage can be very costly, not only financially but also in terms of disruption and disillusion among those working on the project who, by this time, will be mainly in industry.

In summary, the processing of R&D projects in MoD involves their passage through a series of stages, and their appearance before a number of committees – each with different perspectives – such that before each milestone may be passed certain criteria must be satisfied. Projects may be stopped at any stage, and the budgetary pressure upon them increases as they move past the milestones; money is being spent all the time that the project is under consideration, but the rate of spending goes up at each stage. Uncertainties about costs, timing, technical feasibility, collaborative prospects, sales potential, and the strategic environment will, hopefully, have been reduced at each stage, and the availability of the necessary R&D facilities will also have been confirmed. Throughout the process there is a continuous interplay between staff from the different elements within the Ministry – from the Defence Staff, the Office of Management and Budget, the Chief Scientific Adviser's Staff, and the Procurement Executive – to try to ensure that all relevant perspectives on the project are represented.

3.2   THE PERFORMANCE OF THE R&D

So far we have discussed the organization of military R&D mainly from the viewpoint of policy-makers in Whitehall. The actual R&D itself is not, of course, performed in Whitehall but in the defence research establishments, industry, and the universities. In this chapter we consider the first two of these categories, ignoring the universities because in Britain, unlike the USA, their contribution to military R&D is, in financial terms, negligible (see Fig. 2.4).

3.2.1 The defence research establishments

These fall under the direction of the Controller of Establishments, Research, and Nuclear Programmes (CERN), within the Procurement Executive, and their programme is monitored by the Defence Research Committee, chaired by the Chief Scientific Advisor (CSA). The establishments are listed in Fig. 3.4, together with an indication of their work. The current set of establishments is the product of a lengthy process of rationalization which, as we saw in Chapter 2, has been associated with a sharp drop in the number of staff employed.

A second tendency which has underlain the reform of the establishments has been a desire to organize them in such a way that some would become responsible for R&D of military systems as a whole, while others would become more clearly seen as foci for different areas of technological expertise, in both cases on an inter-Service basis. Thus, under reforms announced in March 1984, the vehicles establishment at Chertsey and the gunnery one at Fort Halstead, Kent, became a single land systems research establishment. The Admiralty's underwater weapons, surface weapons, and marine technology establishments became a single sea systems group. Aldermaston and Porton Down meanwhile continued as the centres of technological expertise on nuclear weapons and chemical defence, respectively; the Royal Aircraft Establishment, Farnborough, as the main air systems establishment; and the Royal Signals and Radar Establishment at Malvern continued as an internationally-renowned centre in electronics, electro-optics, sensors, lasers, electronic warfare, and computing. As *The Economist* noted, however, there was less to these changes than met the eye. Few establishments would actually close: the process was essentially one of re-grouping.[1]

A further important tendency has been the transition from, in the early post-war years, doing nearly all military R&D intramurally to a situation today in which two-thirds of all military R&D is done in industry. This transition has largely resulted from the transfer of nearly all the final, full-scale design and development work to industry. Thus whereas, say, RSRE Malvern used to design radars, now this work is almost entirely done in industry. Torpedo development, which as late as 1977 still continued at the Admiralty Underwater Weapons Establishment, has now been transferred to industry. Apart from some aspects of ship design, certain areas of conventional explosives, and some areas of army vehicles work, the final stages of almost all military design and development work now lie firmly with industry, acting as a contractor to the Ministry of Defence through the Procurement Executive. The research establishments' role, in contrast, has reverted to running certain expensive central facilities, performing longer-range research, supporting industrial contractors as required, and evaluating equipment. While some of them, notably RAE (Farnborough) and RSRE (Malvern) do some work on a

There are seven establishments occupying some 20 major sites and a number of minor ones. The main locations of the establishments, together with their primary activities, are as follows:

Royal Aircraft Establishment (RAE)

Farnborough and Bedford — Research covering all aerospace activities including airframes, engines, weapons and systems. Facilities include airfields, ranges and wind tunnels.

Royal Armament Research and Development Establishment (RARDE)

Fort Halstead and Chertsey — Land systems research, including guns and their ammunition, combat and logistic vehicles, engineering and bridging equipment.

Admiralty Research Establishment (ARE)

Portsdown, Portland and Teddington — Sea systems research, including marine technology, weapons and sensors for above and below surface warfare.

Chemical Defence Establishment (CDE)

Porton Down — Research into defence against the threat of chemical and microbiological attack, including physical protection and medical treatment.

Royal Signals and Radar Establishment (RSRE)

Malvern — Radar communications, night vision and other research both applied and fundamental in electronics.

Atomic Weapons Research Establishment (AWRE)

Aldermaston — Research, development and some production aspects of atomic weapons.

Aeroplane and Armament Experimental Establishment (A&AEE)

Boscombe Down — Aeroplane and armament testing, particularly trials for aircraft clearance, equipment evaluation and aerial delivery.

Fig. 3.4. MoD research establishments. (Source: Ministry of Defence (1985). *Statement on the defence estimates 1985*, Cmnd. 9430-I, p. 41. HMSO, London.)

customer–contractor basis for other government departments, most of the work done in the defence research establishments is funded directly by the MoD.

3.2.2 *Industry's role*

Exactly how industry is organized to perform its functions in relation to military R&D is not altogether clear. No systematic data are available on the allocation of qualified scientists and engineers to military R&D in British industry, nor of the commitment made by firms to R&D performed with their own funds (so-called 'Private Ventures'). The systematic data that are available give merely a broad indication of the scale of contract work which firms do for MoD, but these aggregate R&D *and* production. Figure 3.5 lists the UK-based contractors that were paid £5 million or more by MoD for equipment in 1983–84. Many of these work almost entirely in the military field. Others work in both the civil and military fields, but often through separate internal divisions with distinct design and development facilities.

In considering Fig. 3.5 we should first note the major role in defence equipment procurement that is played by the *Royal Ordnance Factories*. In the past they have not done any significant R&D, but this may change under current government plans to privatize the Factories. As part of the preparation for privatization, the 1984 reforms of the defence research establishments, described above, involved the transfer of some 600–700 staff at the rocket research establishments at Westcott, Buckinghamshire and Waltham Abbey, Hertfordshire, to become the foundation of an R&D facility for the embryonic Royal Ordnance Factories company. It is also intended that the tank design and development work done at the former Motor Vehicles and Engineering Establishments at Chobham, Surrey, and at the Royal Armament R&D Establishment at Fort Halstead, Kent, will be transferred to the ROF company. This will leave the restructured RARDE at Fort Halstead working mainly on conventional warheads.

The main performers of military design and development work in private industry and public corporations are likely to be those companies which are the recipients of the largest MoD contracts. Thus British Aerospace, a public corporation, has been involved in the development of the Harrier/Jaguar replacement, the Nimrod maritime patrol and airborne early warning aircraft, and the Tornado multi-role combat aircraft. Its Dynamics Group is a specialist guided-missile manufacturer, having developed and produced, among others, the Rapier, Sea Draft, and Swingfire systems. It has the capability to design and build cruise missiles comparable to current US models. Private sector involvement in R&D is particularly strong in electronics. For example, GEC–Marconi Electronics has claimed that it is involved in every major defence project in the UK for all three Services and in a number of European and American-linked multinational projects.

Finally, in considering industry's role in military R&D, we should note that the prime mechanism for formalizing that role is the *contract* placed by MoD

Over £100 million

British Aerospace plc (Aircraft)
British Aerospace plc (Dynamics)
British Shipbuilders
Ferranti plc
The General Electric Co plc
The Plessey Co Ltd

Racal Electronics plc
Rolls Royce Ltd
Royal Ordnance Factories
Thorn-EMI plc
Westland plc

£50–100 million

Austin Rover Group Ltd
Dowty Group plc
Hunting Associated Industries plc

Philips Electronic &
   Associated Industries Ltd

£25–50 million

General Motors Ltd
Lucas Industries plc
Marshall of Cambridge
   (Engineering) Ltd
Pilkington Bros plc

Short Bros Ltd
Smiths Industries plc
United Scientific Holdings plc
Vickers plc

£10–25 million

Acrow plc
British Electric Traction Co plc
BTR plc
Cable & Wireless plc
Cossor Electronics Ltd
Cambridge Electronics Industries inc
Dunlop Holdings plc
Flight Refuelling (Holdings) plc
Guest Keen and Nettlefolds plc
Harland & Wolff Ltd
Hawker Siddeley Group plc

ICL plc
ITM Offshore
Northern Engineering Industries plc
Oerlikon Buerle Holdings Ltd
Remploy Ltd
Singer Co (UK) Ltd
STC plc
The Throgmorton Trust plc
UKAEA
The Weir Group Ltd

£5–10 million

BICC plc
B Thompson Ltd
Englehard Industries Ltd
Ferguson Industrial Holdings plc
Goodyear Tyre & Rubber Co
Gresham Lion plc
Louis Newmark plc
MacTaggart Scott (Holdings) Ltd
Portsmouth Aviation Ltd
Rank Organisation plc
RCA Ltd
RFD Ltd

Saft (UK) Ltd
Schlumberger Measurement &
   Control Ltd
Siemens Ltd
S Pearson & Son plc
Systems Designers International plc
Thomas Tilling plc
Vantona Group plc
Western Scientific Instruments Ltd
Wilkinson Sword Group Ltd
Yarrow plc

Note: Within each financial bracket, contractors are listed in alphabetical order.

Fig. 3.5. UK-based MoD contractors paid £5 million or more by MoD for equipment in 1983–84. (Source: Ministry of Defence (1985). *Statement on the defence estimates 1985*, Cmnd. 9430-I, p. 68. HMSO, London.)

with a firm which, in the case of major projects, subcontracts to other firms. Contracts can be of several types.[2] The most important follow.

(*a*) *Fixed price contracts* – in which a firm price for the work is agreed at the outset.

(*b*) *Contracts at a price to be agreed* – which may be used when, for example, it is important to get the work underway urgently.

(*c*) *Cost reimbursement contracts* – which are often used in conditions of great uncertainty over the complexity of the project, and which may be:

(1) 'Cost-plus' contracts, which may be further subdivided into 'cost-plus fixed fee' and 'cost-plus percentage', and in which the contractor is reimbursed for all approved costs that have been incurred and, in addition, receives a sum for profit; or

(2) 'Cost-plus-incentive' contracts – which attempt to control the tendency for costs to rise under cost-plus arrangements by introducing an element of incentive into the contract, so that the lower the final costs, the higher the contractor's profit.

Figures provided by the Ministry of Defence show, that in 1983–84, 77 per cent in number, or 38 per cent in value, of all MoD HQ contracts were placed competitively; 13 per cent in number, of 47 per cent in value, were placed non-competitively but under types of contract which implied some risk to the contractor; and 10 per cent in number, or 15 per cent in value, were placed non-competitively and at no risk to the contractor.[3]

4 Policy issues

4.1 INTRODUCTION: IMPERIAL HERITAGE

Britain's commitment to military R&D is, and has long been, among the highest in the world, although it remains an order of magnitude smaller than that of the superpowers. As we have seen, it consumes an extremely high percentage, by international standards, of government annual expenditure on R&D, and of the defence budget itself. As late as 1960, Britain was still aiming at a high degree of autonomy in the development and assembly of military aircraft, tanks, ships, and ordnance, and had the basis of a 2000-mile-range missile programme, together with a wide variety of shorter-range missiles. Even today, the scope of British military R&D remains great, as the list of major programmes authorized for full development or production between May 1979 and 1983 shows (see Table 4.1). The price for staying in the race with this range of equipment is high.

The scale of this enterprise is a legacy of Britain's imperial past. Britain was one of the world's Great Powers up to the Second World War, although by the end of that war it was clear that two Superpowers had emerged.

But old habits died hard. Despite the change in global status, Britain proceeded fairly automatically with an atomic bomb programme[1] and with a host of other research projects, including those which formed the basis of the V-bomber programme. The range of Britain's defence commitment remained global, and even today, with the Empire (and later the East of Suez commitment) gone, and with them the need to defend more than a handful of farflung places, the capability required by Britain's armed forces is still greater than that of any other country in the world, except the USA, the USSR, and, possibly, France. Thus, Britain maintains independent strategic and theatre nuclear forces which are committed to NATO; provides for the direct defence of the United Kingdom; maintains a major army and air force contribution on the European mainland; and deploys a major naval force in the Eastern Atlantic and the Channel. Since the spring of 1982, there has been a substantial military commitment to the Falkland Islands in addition to forces stationed in other colonies, and there continues to be a requirement for an additional limited capability for operations outside the NATO area.

This is the context of Britain's current military R&D portfolio. As a former Great Power and former leader in the development and production of weapons, Britain has inherited a large and sophisticated capability in military R&D, and a tradition of independent and self-sufficient development and production of the weapons needed to support her global role. The passage of time has eroded both

Table 4.1. Major programmes authorized for full development or production May 1979 to 1983

New patrol submarine
Heavyweight torpedo
Improvements in Sea Wolf (naval surface-to-air missile)
Sting Ray torpedo
Electronic support measures for submarines
Sea Eagle anti-ship missile
Skynet IV communications satellite
Challenger and main battle tank improvements (including thermal imaging)
Tracked Rapier (mobile ground-to-air missile)
Towed Rapier improvements
Improvements in blowpipe (man-portable surface-to-air missile)
Battlefield artillery target engagement system
Wavell command and control system
Infantry combat vehicle
Improved tank ammunition
Multiple rocket launch system
Harrier GR5 (AV8B) (improved version of vertical take-off and landing aircraft)
Comprehensive improvement of army and RAF Rapier
Wide-bodied tanker/freighter aircraft
Electronic countermeasures equipment for surface ships

Source: Ministry of Defence (1982 and 1983). *Statement on the defence estimates 1982*, Cmnd. 8529-I, p. 11. HMSO, London; *Statement on the defence estimates 1983*, Cmnd. 8951-I, p. 10. HMSO, London.

the global role and the R&D and production infrastructure, but at significantly different rates, so that the infrastructure is now considered by some commentators to be oversized relative to Britain's world status.

At least seven counterarguments can be advanced against this view.

1. As Britain has the facilities (capital equipment, staff, etc.) for a major military R&D effort, it should exploit them fully.

2. A rundown of military R&D could lead to the break up of highly specialized military equipment research, design, and development teams.

3. A reduction of indigenous military R&D and production capability would damage the balance of payments because necessary equipment would have to be purchased from abroad.

4. Military R&D is valuable for the spin-off into the civil sector which it affords.

5. Self-sufficiency in vital military equipment is important in the event of emergency since, as the Suez experience demonstrated, overseas suppliers cannot always be depended upon.

6. Britain has a technological lead in certain categories of weapons for which there is overseas demand, and there is a belief in Whitehall that foreign military sales offer both a financial and a diplomatic asset.

7. The benefit of redirecting R&D effort away from military projects is not as great as is commonly supposed, although it is an area which needs more thorough investigation.

We will return to the first and last of these arguments later, when we consider the question of redeployment (Para. 5.1). Likewise, the fourth argument will be discussed in Para. 4.4, the fifth in Para. 4.3.1, and the sixth in Para. 4.6. On the second point, we would observe that we have seen no clear evidence – with the one exception of an exodus to Canada and South Africa of staff who had been working on the TSR-2 project until its cancellation – of the wholesale loss to Britain of military R&D staff when defence commitments have been cut in the past, although it is less clear how those staff were subsequently employed. As for the balance-of-payments argument, what needs to be demonstrated is that the cost of British-made weapons is sufficiently close to the cost of imported weapons that an undue price is not being paid to protect the balance of payments.

4.2   ABANDONMENT OF THE NUCLEAR ROLE?

One way to cut sharply the requirement for such a substantial military R&D effort would be to abandon one of Britain's defence roles. The only one whose proposed abandonment would win significant political support appears to be the nuclear one, although it should be noted that decisions over nuclear weapons tend in any case to attract more public interest than decisions over conventional forces, not least when, as with the replacement of Polaris by Trident, such a considerable increase in absolute striking power is entailed. Without pre-judging the desirability, or even the likelihood, of this step, we can consider the effect it would have upon Britain's overall military R&D effort.

The question of what technically is involved in British nuclear disarmament has not until recently been studied in great depth.[2] It seems clear, though, that it would not be possible to engage in nuclear disarmament in a way that could be thoroughly verified by technical means. The size of the nuclear stockpile (including tactical as well as strategic weapons), the difficulties of monitoring its dismantling, and the obligation to protect nuclear secrets obtained from the USA militate against ideal solutions to this problem. Such a policy would, however, initially generate a lot of work for Aldermaston and related establishments in respect both of the actual dismantling of the weapons and the requirement to demonstrate that the dismantling had occurred. But this work would probably be only for a short period and would inevitably leave open the question of the future of the approximately 8500 staff (in all categories) employed on nuclear and nuclear-related R&D. We return later (Para. 5.1) to the possibility of conversion of defence research establishments to civil purposes. It is sufficient here to observe first, that these are not new problems for Aldermaston;[3] second, that another possible option might be for Aldermaston to redeploy into advanced conventional weapons; third, that, according to the Institution of Professional Civil Servants (IPCS),[4] about 80 per cent of the capital equipment at present used in the nuclear R&D field could be directly transferred to civil use – though how they arrived at this figure is unclear; and, finally, that (again according to the IPCS) the number of qualified scientists and engineers and other R&D staff who would be released by a nuclear disarmament programme is likely to be measured in thousands, rather than tens of thousands, and less if there were a parallel expansion in non-nuclear defence work.

4.3 CONSEQUENCES OF OVER-SIZED MILITARY R&D EFFORT

With the leading Opposition party committed to a policy of British nuclear disarmament, the implications of such a move for R&D personnel and facilities should certainly be considered seriously, both inside and outside official circles. However, more immediate problems are already with us. The scale of the British military R&D effort, like the scale of total defence spending, is:

(*a*) Leading to budgetary problems;

(*b*) Fuelling heightened interest in the more efficient use of industry;

(*c*) Leading to added support for international collaboration;

(*d*) Causing concern about its wider economic effects;

(*e*) Increasing the pressure for arms sales.

We consider each of these in turn in the remainder of Para. 4.3 and in Paras. 4.4–4.6, before concluding Chapter 4 with some discussion of the ethics of military R&D.

4.3.1 Budgetary problems

After the post-war economic boom, the economies of most Western nations have suffered from inflation and low economic growth. The cost to those NATO countries that have honoured their 1977 agreement to increase defence spending annually by 3 per cent in real terms until 1985–86 has been sacrifice in other parts of the public sector. In Britain it is clear that the Defence budget will not continue beyond 1986 at the rate of real growth which it has enjoyed in recent years. Yet, at present, the proportion of the defence budget that is spent on research, development, and procurement continues to increase, as the real costs

Table 4.2. Growth in real costs of successive generations of equipment

Lynx Mk 2 helicopter	cost	2.5x	Wasp Mk 1 helicopter
Type 22 Frigate	cost	3x	Leander class frigate
Harrier GR1	cost	4x	Hunter F6

Source: Ministry of Defence (1981). *Statement on the defence estimates 1981*, Cmnd. 8212-I, p. 45. HMSO, London.

of successive generations of equipment grow (see Table 4.2 and Fig. 2.1 for the R&D component of this trend).

This outcome arises partly from an understandable determination to equip the Armed Forces with as advanced weapons as possible, and partly from the growing sophistication of military equipment which, in turn, is related to shorter production runs and therefore fewer items across which to defray the R&D costs. At the same time, critics have characterized the process of incremental innovation along well-established technological trajectories as leading to baroque, and ultimately cost-*IN*effective weapons systems.[5] The Ministry of Defence has itself referred to the need to avoid over-elaboration in weapons requirements.

There are a number of areas where our equipment could be simpler. . . . We must consciously limit the extent to which we exploit all the benefits of new technology, regardless of whether they are necessary or not.[6]

One measure which has recently been adopted to this end is closer involvement of industry in forward equipment planning. Firms are now to be consulted at the earliest possible stage in the definition of an operational requirement, that is, prior to the formulation of Staff Targets. The Staff Targets themselves are to be shorter and simpler, stating the problem and objectives but leaving the solution unspecified. This is known as the 'Cardinal Points' approach.

More selectivity. The government has also called for a more selective approach to R&D.

It is no longer open to Government to maintain the current level of R&D defence funding across the present wide field. Selections will have to be made, and we shall need to consider reducing the range of our defence industrial capabilities and concentrating on a more limited range of weapons technologies.[7]

We consider this conclusion to be inescapable. It has consequences for the organization of military R&D (see Para. 4.3.2) and for international collaboration (Para. 4.3.3), both of which areas offer scope for improvement.

If costs continue to outstrip resources, the consequence will be compromises, skimping, delays, and uncertainties about programmes. Paradoxically, measures such as these, which are designed to try to save money, can actually result in cost over-runs. If, as happened with the Chevaline improvement to the Polaris missiles, money is committed in short bursts (so-called 'trickle feeding'), firms will not commit themselves wholeheartedly to the project, nor put adequate numbers of staff on to the job; the programme will consequently take longer and cost more than if it were dealt with more briskly.

Less autonomy. A further consequence of selectivity must be a less nationally autonomous approach to the development and production of weapons, and a greater emphasis on both international collaboration (see Para. 4.3.3) and imports. Importing would save on direct R&D costs, though no doubt a proportion of the supplier's R&D costs would be included in the price of the equipment. It would also offer the benefit of lower prices resulting from economies of scale in production, as well as contributing to the standardization of equipment within NATO.

The disadvantages would be:

Increased risk of interruption of supply;

Undesirability of dependence on foreign sources for certain equipment;

Loss of capacity in 'leading edge' technologies.

The first of these (which is well illustrated by the problems of maintaining the Polaris missiles after their removal from service in the USA) would depend upon the state of the supply side of the market (which, as suppliers proliferate, is on the whole increasingly becoming a buyer's market), and the state of Alliance relations. The second, and related, disadvantage derives from the argument that some technologies are unlikely to be available for purchase from abroad (nuclear weapons being the best example), and that others (such as encryption or electronic countermeasures) should not be as a matter of prudence. As for the third disadvantage, particularly in the case of arms imports from the USA, it is rarely possible to buy component technologies; one is offered a total system or nothing. One could not, therefore, easily put British microelectronics into essentially American systems, and this could result in reduced demand for British microelectronics, and associated loss of industrial capability, if American systems were substituted for British.

These objections, together with such others as the balance-of-payments argument, make greater dependence on imported weapons unattractive to politicians (and this despite the costs of maintaining a domestic arms industry) and, of course, to the defence industries. One quarter from which wholesale opposition would not be expected is, however, the Armed Forces. They often, though not invariably, give serious consideration to buying off-the-shelf from abroad, provided that what is available offers something very close to the desired specification. They see this approach as offering cheapness (at least for prime costs), more rapid availability, tried equipment (the bugs having been blown out elsewhere), and, as already mentioned, a contribution to interoperability and standardization. They are, however, well able to insist upon

British-made equipment if an appropriate alternative cannot be bought off-the-shelf abroad.

4.3.2 *Making better use of industry*

Another possible escape route from rising equipment prices might be to try to cut the costs of military R&D by putting more of it into industry, while simultaneously making industry more competitive. As we have seen (Para. 3.2.1), there has already been a marked shift over the past two decades in the distribution of military design and development work between government research establishments and industry. Associated with this shift have been substantial cuts in the number of defence research establishments and their staff (Chapter 2). Industry, moreover, not only carries more of the responsibility for design and development than it used to, but is also involved at an earlier stage in the process.

Has this policy been pressed to its limits? In February 1984, a contributor to *The Economist* made the radical proposal that all military R&D (with the exception of nuclear research) should either be transferred to industry or managed by it.[8] The remaining government research establishments would be confined to testing, proving, and acceptance work. The same source also recommended the opening up of the defence contracts business claiming that, at present, up to 70 per cent (in value) of contracts are single-source or otherwise non-competitive (but cf. the rather differently based figures cited in Para. 3.2.2), and over 60 per cent go to 10 large companies. If, in addition, firms were required to put up, say, 30 per cent of the cost of an R&D project, this could bring much-needed discipline.

There are certainly grounds for believing that the current arrangements for financing military R&D and equipment procurement warrant review. Budget-seeking bureaucracies have been said to stress the need to 'match the threat', without adequate regard to cost. Weapons can then acquire the characteristics of 'gold plating', complexity, and a desire to outstrip the equipment of friendly as well as rival armed forces. It would follow that technical sophistication could then be pursued for the satisfaction it gives to scientists, producers, and operators of weapons, more or less regardless of cost. Moreover, vote-sensitive governments will always be alive to the political dangers of *not* buying British. From all this, state support for British defence contractors has, it has been argued, created non-competitive or, at best, oligopolistic suppliers who in the past at least (for things are now changing) have been further protected by a system of cost-plus contracts which ensured that they recovered their actual expenditures almost regardless of the level.[9]

More competition. The thrust of these arguments is that military R&D should be progressively moved into industry, and that industry itself should accept more of the risks and operate in a more competitive environment.

While the latter point may have some validity, especially if, as the

Government have claimed, cost savings from competition amount to over 30 per cent,[10] we question the wisdom of reducing the defence research establishments to the rump which these proposals would entail. They have already undergone substantial contraction, and almost all non-nuclear design and development work has been transferred to industry. If the tasks of concept formulation, and generation and vetting of specifications, were also transferred to industry, then indeed the establishments could be further cut until all that remained would be testing, proving, and acceptance work. However, in yielding these activities to industry, government would also be weakening its ability to act as an informed customer for future military equipment; headquarters staff alone, without the back-up provided by technical staff in the research establishments, could not be expected to perform this task as well as it is done at present.

Other difficulties. But even if these transfers were effected, other difficulties would remain.

(a) One is the loss of human resources incurred in dispersing the skilled teams currently assembled in the defence research establishments. We have already expressed some scepticism about the significance of this argument, but we would concede that one establishment in particular, RSRE (Malvern), has an outstanding reputation as a centre of excellence, and that it already devotes a significant part of its total effort to civil R&D under contract to the Department of Trade and Industry. Much the same could be said of the space research at RAE (Farnborough). We would also agree that the implications of a transfer to industry would be particularly acute for longer-range research, and that this problem would require further thought.

(b) A second difficulty concerns the possible effect of increased government R&D spending in industry on industry's *own* commitment to R&D. This is a much under-examined question; the evidence on it is conflicting, and may well vary by industry sector. Nevertheless, there is some evidence from the United States (see Para. 4.4.2) that, far from encouraging industry to spend more of its own dollars on R&D, the effect of federal government spending on R&D in industry is simply to depress the level of company-financed R&D. We have already expressed concern about the level of spending on R&D by British manufacturing industry; any further depression would hardly be helpful.

(c) A third difficulty concerns industry's reaction to a more competitive environment. The present government is already committed to increasing the proportion of contracts that are awarded competitively. The example of the MCV 80 armoured infantry carrier is a case in point. Contrary to precedent, the firm that did the development work, GKN Sankey, were then given only a limited production contract. Other firms were invited to tender for the rest of the production run. This approach has raised doubts within firms about the wisdom of too heavy a commitment to military R&D if the more profitable work, the production contract, may go to a competitor. There is, we believe, an element of special pleading here: the solution to the problem of whether it is worth a firm engaging in military R&D surely lies in the terms of the contract, and it may well be that in the new climate, government may have to pay firms more to do R&D if it wishes simultaneously to create uncertainty about whether a production contract will follow. This point runs counter to the proposal, cited earlier, that industry should put up 30 per cent of the cost of R&D, and raises a doubt about the savings to be obtained by this strategy. A further problem with a policy of increased competitiveness is that in some sectors of the defence equipment business the market is so small and specialized that there are simply not enough *British* firms operating to enable significant *domestic* competition at prime contractor level to take place. (Things may well be different at subcontractor level.) If the MoD wishes to maintain an indigenous capacity in these sectors, it must take care to balance that objective against the one of increasing competition. Otherwise, a competiveness policy may turn into an imports policy, and while the pursuit of *international* competition may be a valid goal, it should not be introduced by default. Ironically, the limited base for domestic competition is to a considerable extent the result of previous government policies as, for example, over the concentration of the aircraft industry into ever fewer firms (although it is also fair to note that the British aircraft market could not now support several firms large enough to compete internationally). A final problem with a policy of increased competitiveness is that it may not take adequate account of the problems of what Edmonds had termed 'defence oriented companies', that is, companies whose elasticity of substitution of output is such that they must totally change the nature of their business if demand from the MoD collapses.[11]

4.3.3 International collaboration and standardization

Another route to economies in military development and production lies through the processes of equipment standardization and international collaboration. Standardization involves the deliberate purchase of the same equipment by more than one NATO country. Collaboration involves agreement to collaborate on R&D and/or production. The link between the two is that collaboration on R&D increases the chance of common production programmes; standardization, even without prior collaboration, can nevertheless be beneficial not merely because of the operational advantages of standardized equipment, but also because it offers the chance of defraying R&D costs over longer production runs.

Lack of standardization within NATO was estimated in 1975 to be costing over $10 billion per year in duplicated R&D and production costs, amongst other things.[12] More recently, 11 firms in seven NATO countries were reported to be working on antitank weapons; 18 in seven countries on ground-to-air weapons; eight firms in six countries on air-to-air weapons; 16 in seven countries on air-to-ground weapons; and 10 firms in seven countries on ship-to-ship missiles.[13]

While excessive standardization may so ease the task of devising countermeasures that it becomes militarily counterproductive, such a dispersion of effort supports Callaghan's argument that NATO is engaged on a process of 'structural disarmament'.[14] At the same time, there is no shortage of barriers to standardization. These include:[15]

Domestic politics (including the protection of employment and industrial capacities).

Differences in military requirements.

The sheer size, relative to Europe, of the US economy which has made it unnecessary for the US to pool resources with other nations.

The concomitant US technological equality or lead in most areas of military equipment, which makes equity in trade difficult to achieve.

Sheer diversity of weapons *within* as well as *between* nations, with air forces and navies, for example, duplicating R&D, equipment, and training.

Arms sales to traditional customers.

Prejudice.

Renewed impetus for collaboration. Although progress towards standardization is likely, therefore, to be slow, collaboration has recently received a political boost, after having fallen back in prominence in the early-1980s from where it was in the mid-1970s. Administratively, the post of Chief of Defence Equipment Collaboration was created in March 1985 to concentrate full-time on equipment collaboration (cf. Fig. 3.2). The then incumbent of the post of Chief of Defence Procurement was moved into the new post.

There is, of course, already considerable experience of collaboration (see Fig. 4.1), in the form of joint projects such as Tornado, or a division of labour under which one country develops one weapon in a 'family of weapons' (such as short- or medium-range advanced air-to-air missiles) while another develops another. Much of this experience has been encouraged under NATO and Western European Union auspices, and initiated after direct dealings between the countries concerned. Most recently, and with some prompting from the British government, the Independent European Programme Group (IEPG), which comprises the European members of NATO, including France, has provided a forum for discussions about, *inter alia*, a European Fighter Aircraft and a European response to the American initiative over the development of emerging conventional technologies (ET).

The emerging technologies draw upon advanced work in microelectronics, sensors, materials, and command, control, and communications, and offer the possibilities of real-time, long-range, acquisition of fixed and mobile targets, and the capacity to launch highly lethal attacks by, for example, terminally-guided 'submunitions' which could distinguish between trucks and tanks in the final moments of attack. ET, in short, offers the prospect of much enhanced *conventional* defence by multiplying the technical capacity of existing NATO

Project	Participating Countries
In service	
Naval Equipment:	
PARIS Sonar	UK/FR/NL
Land Equipment:	
FH70 Howitzer	UK/GE/IT
Scorpion Reconnaissance Vehicle	UK/BE
Aircraft:	
Jaguar	UK/FR
Tornado	UK/GE/IT
Lynx)	
Gazelle)	UK/FR
Puma)	
Missiles:	
Martel (Air-to-Surface)	UK/FR
Milan (Anti-Tank)	UK/FR/GE
Sidewinder (Air-to-Air)	UK/GE/IT/NO
Other Equipment:	
Midge Drone	UK/CA/GE
In Development or earlier Study Phases	
Naval Equipment:	
NATO Frigate Replacement (NFR 90)	UK/US/NL/FR/CA/SP/GE/IT
Sea Gnat Decoy System	UK/DE/US
Land Equipment:	
SP70 Howitzer	UK/GE/IT
Multiple-Launch Rocket System Phase I	UK/FR/GE/IT/US
Multiple-Launch Rocket System Phase III	UK/FR/GE/US
Aircraft:	
Harrier GR5	UK/US
Naval ASW Helicopter (EH101)	UK/IT
European Fighter Aircraft	UK/FR/GE/IT/SP
Missiles:	
Short-Range Anti-Radar Missile	UK/US/BE/GE/CA/NL/IT
Long-Range Stand-Off Missile	UK/US/GE
Milan Improvements	UK/FR/GE
TRIGAT (Anti-Tank)	UK/FR/GE
ASRAAM (Air-to-Air)	UK/GE/NO
Other Equipment:	
Midge Post-Design Services	UK/FR/GE
Note:	
BE = Belgium; CA = Canada; DE = Denmark; FR = France; GE = Federal Republic of Germany; IT = Italy; NL = Netherlands; NO = Norway; SP = Spain.	

Fig. 4.1. UK collaboration. (Source: Ministry of Defence (1985). *Statement on the defence estimates 1985*, Cmnd. 9430-I, p. 17, HMSO, London.)

forces. This prospect has become important in the face of growing unrest in Europe about NATO's current dependence upon the early use of nuclear weapons, and has gone hand in hand with the introduction of new strategic approaches, such as the US Army's Air Land Battle doctrine, and NATO's Follow-on Forces Attack doctrine.

These doctrines are themselves controversial. What matters for our purposes is that from a broadly-based desire for better conventional defence, coupled with the potentially very high costs or 'ET', and with European concern that unless they collaborate over these new technologies, American firms will sweep the board, has come renewed political interest in collaboration in military R&D. Agreement was reached in May 1984 to proceed with 11 projects under the ET initiative in the fields of surveillance, communications, and precision artillery, while in November 1984 the IEPG countries agreed to work co-operatively on a new heavy battle tank for the late-1990s, a surface-to-air missile, and a military transport plane.

These are all constructive steps. Moreover, to the extent that collaboration works best when conducted between equal partners, success should be easier today than in the past. Whether these steps will be persisted with, as projects move through development to production, the risks rising in parallel, remains to be seen. The record of past collaborative projects is sufficiently mixed[16] for there to be plenty of scope for those who wish to question whether, from a British perspective at least, collaboration with European partners is worth the effort. Collaborative projects suffer from delays, high administrative overheads, and complex designs to accommodate the differing operational requirements of the collaborating partners. Ultimately, each participant has only a fraction of the total production. On the other hand, wholly national projects, it can be argued, though suffering from shorter production runs and higher national R&D costs, may offer greater speed of development plus all the production work. Just such arguments have impeded progress with the European fighter project, although the project now looks likely to proceed with four participating nations. Nevertheless, there is evidence that, given adequate political commitment and good management, and allowing for possible variations between sectors, collaboration can play a big part in containing individual nations' armaments bills, just as in one sector of the civil aircraft market, Airbus Industrie has shown how Europeans can beat the Americans at their own game in a highly competitive world market. Hartley[17] has suggested that collaborative projects can add 30 per cent to the R&D costs, and production inefficiencies in the region of an additional 1–10 per cent for a given output, when compared with the estimated cost of the same project, were it performed wholly nationally. Nonetheless, once those total costs are shared between two partners, each will save at least 35 per cent on production costs, compared with the cost of an independent project. All this, of course, says nothing about other possible benefits from collaboration, such as political, diplomatic, and technological ones.

4.4   ECONOMIC EFFECTS OF MILITARY R&D

In addition to general arguments about the cost of modern military R&D programmes, they have a further consequence which we must examine. We refer to the wider economic effects of the investment of resources in military R&D.

It is important to be clear that the primary purpose of Britain's investment in military R&D is to meet Britain's defence requirements. From the perspective of the Ministry of Defence, all else is secondary – and rightly so. But, resources of people and money being constrained, we (and, indeed, the government and the nation) must be concerned with the general economic consequences of this investment. We need to be sure that it is being managed in such a way as to maximize its benefits and minimize its costs to the economy. The subject is, however, complex, and empirical evidence is scarce. Our discussion must, therefore, be inconclusive, but we hope that we can indicate various approaches to the question and convey some of its complexities and uncertainties.

Before focusing closely on military R&D, let us first note that criticism of defence spending as a whole on economic grounds has usually been based on the proposition that it drains the economy of capital and people that could have been employed in more economically productive activities. This proposition has then been linked to explanations of Britain's economic decline. On the other hand, defence spending provides a public good in the form of security for which society is willing to pay. Moreover, in 1980–81, defence equipment expenditure directly supported some 240 000 jobs in British industry, and another 190 000 indirectly. Exports of military equipment sustained an additional 140 000 jobs.[18] Defence sales represent some 25 per cent of the output of the British defence equipment industry and amount to some 2.5 per cent of total British exports.[19] For some companies, a high volume of exports is crucial to their survival. Clearly, military R&D contributes to these achievements, although it is not entirely clear on the available evidence, if one looks at the entire sweep of defence equipment exports, that the export successes in value terms (notable exceptions, such as Rapier anti-aircraft missile systems, apart), correspond well with the areas of high R&D intensity.[20]

4.4.1   Spin-off

The central question concerns the opportunity cost of military R&D: it undoubtedly has contributed to the support of a large number of jobs, but at what cost in terms of potentially more productive opportunities? Would the resources currently employed in defence make a greater contribution to jobs, technology, the balance of payments, and, ultimately, to human welfare if they were used elsewhere in the economy? One response to this question is to cite evidence of incidental benefits from military R&D in the form of 'spin-off'.

Wartime experience is often referred to in this context. There is no doubt, for example, that a number of extremely valuable civil technologies ranging from radar to penicillin were brought to maturity in a far shorter period under the intense stimulus of war than they would have been in peacetime conditions. But these conditions *were* exceptional, and cannot (fortunately) be made the norm for military R&D as things are at present. It is not obvious, for example, that the competitive pressure for innovation is actually greater in military technology than it is in many consumer products in international trade.

Nevertheless, there are many examples of military technical developments which have diffused into civil application. Among the better known are nuclear power, liquid crystals, aeroengine propulsion, carbon fibres, and thermal imaging. These, and other successes, have prompted government to facilitate the civil application of the results of military R&D, and a number of initiatives in this respect have been taken since 1979.

One such was the establishment of joint MoD–industry committees to inform industry about the defence research programme, though the House of Commons Defence Committee later found that these had chiefly involved only the major contractors. The then Department of Industry also expressed concern to the Committee that MoD was not doing enough to use its massive public purchasing power to stimulate industrial innovation.[21] Then, in 1983, the Maddock report to the Electronics Economic Development Council argued that too little was being done within the major electronics defence contractors to encourage spin-off.[22] The House of Lords Select Committee on Science and Technology reached a similar conclusion in 1983; it spoke of the creation of a circle of large firms which had prospered through MoD contracts and which kept largely to themselves the R&D associated with this work.[23]

By the end of 1983, the government, prompted by prime ministerial interest, had begun to take steps to deal with the problem of firms 'sitting on' the results of military R&D. It had also launched a scheme which led to the establishment in December 1984 of an organization called Defence Technology Enterprises (DTE). DTE is funded by a consortium of venture capitalists. They are being granted special access to RSRE (Malvern), RAE (Farnborough), and ARE (Portsdown) with a view to identifying commercially applicable ideas which can be cleared on security and commercial sensitivity grounds. They then communicate these ideas to their (much larger number of) associate members, possibly arranging further venture capital to help the projects on their way. The MoD has a share in the profits from DTE and receives a licence fee for all ideas transferred through DTE.

Schemes such as this are welcome. However, from an analytical perspective, 'spin-off' is not, in our view, a very helpful concept for trying to evaluate the impact of military R&D. For one thing, it may seem to imply that all that government research establishments ever transmit to industry is fairly complete inventions. Yet, in the rather analogous case of universities, there is some evidence to suggest that of at least equal importance is the general advice about promising lines and blind alleys (negative spin-off, which helps to avoid the repetition of errors) which university scientists pass on to industrial colleagues.

If the analogy is valid, then we might expect defence research establishment staff similarly to pass on ideas which, while not developed to the point of a concrete invention, may nevertheless, be of great value. If they do not do this, they should be encouraged to do so. However, the key point here is that looking for examples of spin-off will not expose this sort of activity.

A second reason for doubting the analytical usefulness of the concept of spin-off is that any attempt to use it rigorously encounters difficulties in identifying and measuring the output from R&D programmes in a systematic and controlled way. Much of this output will be 'tacit knowledge', that is, it will be about *how* to do things, rather than about concrete inventions. A high proportion of this will involve industry–industry rather than government–industry transfer, and will therefore be doubly hard to identify. Moreover, much spin-off is likely to be of a 'small beer' nature, thus compounding the problems of identifying it.

4.4.2 Structural impact of military R&D

A more useful approach may be to examine the long-term structural impact of military R&D on the economy. No work of this type has yet been done for Britain, but we can illustrate the approach from US studies.

There has been some analysis of the structural effects of US military R&D spending by industry sector, seeking correlations between the level of military R&D spending in a sector and that sector's economic performance as measured by, for example, productivity, market share, or trade balance indicators. If the benefits to the sector outweigh the costs, then sectors with high military R&D spending should perform better than those with little or no military R&D spending, all other things being equal.

In fact, the contrary seems to be the case. There is evidence that US industry has suffered some of its largest market share reductions in industries that are heavily engaged in military contracting, especially aircraft, electronics, and machine tools.[24] The economic benefits of military R&D have not apparently outweighed the costs, by helping to maintain or expand the share of US products in world markets. The Japanese have significantly penetrated markets for electronic memory chips and computer-controlled machine tools, while in the commercial airline market America's dominance remains, but is being challenged to a limited though still significant extent, by Airbus Industrie. (We do not, however, know the extent to which Airbus is being subsidized.) Moreover, a rough, negative relationship has been found between the share of GDP spent on military R&D and the rate of productivity growth among major industrial nations (see Fig. 4.2). The explanation that is offered for this finding is that nearly all military R&D has been focused on specific applications; throughout the last two decades, only about 3 per cent has gone for basic research.

Although these two sets of data are only suggestive, they cast doubt on the proposition that military spending helps civil technology more than it hinders it.

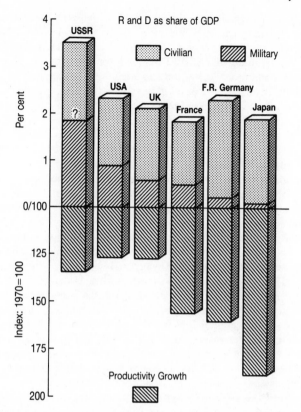

Fig. 4.2. Average military and civilian R&D expenditure as a share of GDP vs. productivity growth in manufacturing industries: selected nations 1970–79. (Source: US National Science Board figures, cited in DeGrasse, R. W. Jr, (1983). *Military expansion economic decline*, Chart 3.4. Council on Economic Priorities, New York.)

Numerous factors influence international competitiveness and production efficiency. Yet the trends of those data support the thesis that the negative effects of military spending on technology outweigh the positive spin-offs. 'The decline in productivity and industrial standards in the US,' commented one Japanese observer, 'is the best argument against the idea that more defence contracts are vital to maintaining "state of the art" efficiency.'[25]

A case study of the US electronics industry extends this argument, though not straightforwardly.[26] The first, unsurprising, finding was that products developed for military use have had few commercial applications. Second, military R&D in semiconductors had generally not produced basic technological innovations. However, the military did contribute to their development, and in a very important way, by recognizing the potential of many of the new electronic innovations and, in effect, heavily supporting the development of the industry through early purchases and through ensuring a steady cash flow which provided a foundation for further civil work. Thus the invention of the transistor in 1947 owed nothing to military interest. But subsequent military support for the infant industry made it possible to develop cheap and reliable transistors. By the late-1950s, when two further key innovations occurred, the military was heavily engaged in electronics research, but none of its projects was instrumental in these discoveries, viz. Texas Instruments' invention of the integrated circuit, and Fairchild's development of the planar process for mass-producing silicon chips. As with the transistor, however, the military market for integrated circuits provided the development and production experience necessary to make these devices commercially attractive. Metal Oxide Semiconductor (MOS) technology and the microprocessor were also initially developed entirely independently from the Pentagon's R&D programmes.

Third, in association with these last developments, a major change took place in the semiconductor industry. During the 1970s, not only did the Pentagon continue to fail to pick the research winners, but it also became decreasingly important as the first user (and, in effect, incubator) of new electronic developments. In the 1970s, civilian demand for semiconductors became sufficiently large that this market, rather than the military, drove technology advancement. The government's share of semiconductor output fell from 36 per cent of the total in 1969 to about 10 per cent in 1978. This was the result not so much of falling government demand as of the explosive growth in the use of semiconductors in industrial and consumer products during that period. Hence, the military acquired a sense of declining control over the direction of the US semiconductor industry, and responsed by launching the Very High Speed Integrated Circuits (VHSIC) programme. This, however, raised fears in other quarters that the result will be to retard commercially applicable technological progress.

Other US work supports this line of argument. Studies of the development in the United States of integrated circuits and numerically controlled machine tools suggest that military domination of these fields led to lines of technological development which diverged markedly from the requirements of civilian markets.[27] Contrast the Japanese experience in these same fields, as reflected in their success in export markets, and particularly in the US market itself.

A further refinement of the US debate has been the attempt to examine possible linkages between the volume of federally-financed R&D and company-financed R&D. (Federally-financed R&D is very largely for defence-related work.) The underlying argument is that even if productivity growth cannot be shown to have followed directly from the investment of federal R&D dollars, perhaps, nevertheless, those dollars stimulated *company*-financed R&D and in that way yielded an economic benefit. The use of dollars as a measure of R&D inputs may, however, have been misleading for various technical reasons. A different indicator of R&D input, namely, employment of scientists and engineers, produces more pessimistic conclusions about the capacity of federal R&D spending to stimulate company spending.[28] In other words, federal R&D spending appears to draw a company's R&D staff away from other, company-financed, work and does not lead to the employment of more R&D staff.

4.4.3 UK studies

We have not come across any comparable work for the UK. Recent work by Soete and Dosi, however, suggests that similar conclusions might be found for the electronics sector.[29] Their study, which was on technology and employment in the electronics industry and not on military R&D as such, shows that this sector is the most R&D-intensive of all British industrial sectors; it has overtaken aerospace since 1975, and accounts now for more than 30 per cent of R&D expenditure in industry. More than half the R&D funds in the sector come from government, and more than 90 per cent of these from the Ministry of Defence. Looked at another way, UK private industry now spends less of its own money on electronics R&D than French industry or any other of its major competitors. British firms have failed to become leaders in most fields of semiconductors or other electronic developments over the past 30 years, with the major exceptions of military electronics and areas within electronic computers and electronic instruments. This also relates to the declining international competitiveness of the UK electronics industry. A trade surplus of £106 million in 1963 was replaced by a deficit of £876 million in 1982; only in electronic capital goods, which are dominated by military electronic equipment, is there still a trade surplus.

Obviously, there is more than one interpretation of such results. This may be a case of military 'distortion' of the sector; alternatively, perhaps the performance of the sector would have been even worse had it not enjoyed military support. We should also beware of generalizing from evidence in only one industry sector. It might well be, for example, that in aeroengines, the British civil sector has benefited enormously from its interaction with military developments. This is a question that should be looked at closely by those concerned with industrial and economic policy, while in the case of the defence sector, we would hope for greater interest in maximizing the use of civil-originated technology, both to conserve military R&D funds and in the interests of using all the government's purchasing power to support socially beneficial projects in British industry.

Just such a mixture of defence and civil interest is exhibited in the Alvey programme on fifth-generation computing. £350 million has been provided for this programme, with a significant contribution from the MoD, together with the Department of Trade and Industry, the Science and Engineering Research Council, and industry. The MoD has also committed research staff to the Alvey projects and advisory panels. Naturally, the Ministry hopes for a good return on its investment. The Alvey work on Intelligent Knowledge Based Systems could, for example, help to alleviate a bottleneck in software design capacity which some foresee as likely to arise from the current rapid uptake of computerized systems into military equipment. For instance, half of the development cost of the Ptarmigan digital trunk communications system that is being adopted by the British Army of the Rhine has been spent on system design and software development. The Type 23 frigate may have more than 200 computers on board. In the decade 1978–88, the number of battle system computers deployed by NATO is expected to rise by 50 to 100 times; data transmission capacity by 10–100 times; software deployed in battle systems by 1000–10 000 times; and so on.[30] The problem faced by MoD is clear enough, therefore, as is the good sense of the joint effort with the Department of Trade and Industry and the Science and Engineering Research Council. What must be watched carefully, however, is any tendency for the MoD interest to sway the overall programme in directions which reduced the prospects for civil applications.

Much weight is being put on the longer-term economic consequences in the current British and European debate about the Strategic Defense Initiative (SDI). European governments are sceptical about the strategic and military benefits of the initiative yet are also concerned that the boost which the SDI may give to a range of advanced technologies will result in a new American technological challenge. We remain unconvinced by the argument that the SDI will produce valuable civil spin-off, and note that the prospects for British industrial involvement in actual production (as distinct from R&D) are even more remote. We incline to the view that if European governments are concerned about a new technological challenge, the way to meet it is by selective encouragement of advanced technologies that are relevant to identified European needs. We do not consider it wise to attach the UK to the SDI on the dubious grounds of spin-off. The French Eureka proposal, and the proposals emanating from the Commission of the European Communities, may not yet be satisfactory either, but they do seem to be more attuned to Europe's needs than is the SDI.

As a final word on the economic effects of military R&D, we offer the thought that most military R&D leads to *product* innovation. Much of the innovation upon which civilian industry depends is, however, in improvement in the manufacturing *process*, not in new product development. It is through process innovation that companies compete over price and quality, thus enabling them to stay in markets that are threatened by, for example, foreign competitors with lower operating costs. To the extent that military development is concerned with products, and not with improving manufacturing processes (and this is not always the case), the scope for benefit to the civilian economy must be expected to be limited. While there are some signs of MoD interest in improving processes (such as in the field of Flexible Manufacturing Technology), there is room for much more to be done.

4.5 THE PERSONNEL FACTOR

Rich though Britain is by world standards in scientists and technologists, we are still heavily constrained in certain areas. At present there is a world-wide shortage of trained people in information technology (IT), and in Britain alone the supply of 5500 IT graduates in 1984–85 will fall short of the estimated demand of 7000, of which a significant fraction is in the defence sector. If the latter were to get more than its proportionate share of the available supply, the consequences might be very serious for the fast growing and economically vital civil IT industry. There is a similar shortfall in this sector in the USA – a most serious potential 'drain' for British specialists.

The percentage of Britain's total stock of *scientists* in the Scientific Civil Service as a whole was between about 4 and 6 per cent during the 1970s, while the percentage of *engineers* was about 3 to 4 per cent.[31] Of these, certain skills were heavily concentrated in the Ministry of Defence. For example, in 1980, 1168 of the civil service total of 1241 classical physicists, and 1340 of the total of 1420 electronic engineers were in MoD, and there were also high concentrations of applied mathematicians, chemists (other than analyticals), computer scientists, engineers of all types, metallurgists, and solid-state physicists. Without comparable figures for industry's commitment to military R&D, and without knowing the quality of these people, it is impossible to say whether the best of the nation's scientists and engineers are being drawn disproportionately into military R&D.

Nevertheless, it is widely accepted among people well-placed to make the judgement that, in most areas of R&D, the demands of military work, and the dynamism which accompanies it, are greater than in the civil sector. That being a popular conception, it would be surprising if it did not lead to preferential selection of military R&D by the brightest and best. Even if this is not the case, such people are always in short supply, and would bring considerable benefits to the civil economy if they were employed there, although, of course, there can be no guarantee that these scientists would be employed on civil R&D if they were not employed on military R&D.

The possibility that an undue proportion of scarce talent is to be found in the military R&D sector would not worry us greatly if there were a high rate of circulation of scientists and engineers between the civil and military sectors. Again, we lack specific data, but all experience of interchange of professional staff suggests that although matters may be improving, the extent of mobility is still fairly limited. The usual problems of pensions, costs of moving, and so on, impede movement between government and industry, or the academic world, as much in this sector as they do in the general case. In addition, there may be special barriers to mobility across the military–civil divide arising from the characteristics of military R&D. Much of it is, after all, subject to security classification. People engaged in such work, whether in government or industry, presumably do so out of a mixture of motives which will often include a desire to work at the frontier of a field and a fair amount of patriotism (see Para. 4.7). Such movement as there is probably remains mostly *within* the 'military R&D community', whose values, standards, and techniques cross the government–industry divide more easily than they cross the civil–military divide. There is certainly a strong ethos in the defence research establishments that encourages the diffusion of ideas into the defence industries, and only secondary concern with diffusion into civil industry. We might also speculate that the nature of the work, and the lack of public appreciation of it, would encourage toilers in the military R&D field to become something of a class apart. This, again, would reinforce the difficulties of transferring ideas from the military realm into the civil economy.

4.6 THE ARMS TRADE

The final consequence of Britain's commitment to military R&D which we wish to discuss is its implications for the arms trade, and especially the desire to recover some of the costs of military R&D through exports. Britain is fifth in the world league table of arms exporters. Arms exports sustain 140 000 jobs and amount to some 2.5 per cent of total British exports. They represent some 25 per cent of the output of the British defence equipment industry, thus enabling R&D costs to be defrayed over longer production runs than would otherwise be the case.

In a heavily armed world, all peace-minded people must view with concern the further spread of weapons, especially highly advanced weapons. On the other hand, it can be argued that the provision of weapons is sometimes in the interests of international security. There is the further argument that if 'we' don't supply the weapons, 'others will', and the others include superpowers, European powers, and industrializing countries like Brazil and India. Hence, it is argued, is it not better for 'us' to supply the weapons, thereby possibly both retaining a moderating influence over their use and cementing better relations with the customer country in the hope that trade, which once followed the flag, may now follow the gun?

This is a double-edged argument. It takes for granted Britain's own moral superiority. And it offers no guarantee that the UK itself will not suffer diplomatically from this trade. Whereas *Realpolitik* imposes a direct obligation on a superpower such as the USA or the USSR to cement alliances with other countries within its bloc by the supply of arms, this obligation does not rest on medium-sized nations such as the UK or France, whose foreign policy interests rest primarily on very close relationships with their regional partners.

Accordingly, we wish to strike a note of caution about the systematic incorporation of an export dimension into decisions about British military R&D and production. The Defence Sales Organization is now closely involved at all levels in decisions about military equipment programmes. It is explicit government policy that export potential should be an important element in the drawing up of equipment specifications. Ministers, including the Prime Minister, travel the world trying to assist the sale of British weapons. This assistance is particularly important at present in view of the effects of world recession in depressing the market for arms exports; by the same token however, the depressed state of the market is a further reason for not banking too heavily upon arms exports. We know that the decisions about specific cases are made after careful interdepartmental consultation, with the Foreign and Commonwealth Office taking a particularly close interest. We understand the desire to recover some of the costs of British military R&D, and to preserve parts of the defence industrial base, by selling its products abroad, but we also consider that Britain's foreign policy interests do not always benefit from strong commitment to this trade. We accept, however, that this is an area in which difficult balances must be struck. In particular, just as we do not wish arms sales to be divorced from foreign policy, so we must also guard against adopting a patronizing air of appearing to know what is in the best interests of other countries; this is not an attitude which Britons welcome when applied by other countries to the UK itself, as has happened, for example, in the case of Anglo-American defence relations on occasion. Moreover, if Britain wishes from time to time to *buy* weapons systems from abroad (Trident being perhaps the most notable current case), then it certainly cannot deem arms *sales* to be automatically immoral.

4.7 THE ETHICAL DIMENSION

To those employed in the relatively closed world of military R&D, it is a profession like any other. The fact that they may be involved in the creation of ever more powerful and effective tools of death or destruction does not obtrude continually into their thoughts. Their regular concerns are technical, or organizational, within the narrow bounds of their particular responsibilities and their proper concern with national security in an imperfect world.

They would say – and many would agree with them – that this is as it should be. The functions they perform are considered vitally necessary by the society in which they live; they are mature, responsible, and patriotic citizens, earning their living in honourable employment; and there can therefore be no suggestion that they should not give of their best in this work, as laid down for them by their superiors.

Nevertheless, many other people outside their profession feel grave concern about its ultimate morality. A strict pacifist, for example, would find the role of the armourer just as unacceptable as that of the soldier; indeed, from that point of view, persons who make and sell warlike weapons are even more to be blamed than those who risk their own lives in using them. Only a small minority of people put strict pacifist principles above all other considerations – but the validity of this point of view as an ideal of individual and collective behaviour is widely held, in Britain as in many other civilized, open societies. Nor is it necessary to be a pacifist in order to have grave misgivings about the moral legitimacy of an escalating arms trade, one of whose driving forces is the genuine enthusiasm of defence scientists and engineers for the design and perfection of ever more sophisticated, 'effective', and, inevitably, costly weapons of death and destruction.

This attitude, which is obviously linked to strong feelings about the dangerous possibilities of nuclear war, on the one hand, and belief in the need to maintain a strong deterrent, on the other, is not irrelevant to this report. The question is raised, for example, whether the sheer activity of military R&D has become a primary cause of the armaments race. Stated thus, of course, this proposition begs as many questions as it asks. Nevertheless, well-informed commentators with inside knowledge of the R&D system, such as Lord Zuckerman, and the late Earl Mountbatten, have publicly expressed the opinion that weapons innovation nowadays is driven more by the technical zeal of the scientists than by the needs of national defence. More recently, with reference to the Strategic Defense Initiative, the Foreign Secretary, Sir Geoffrey Howe, in a speech delivered to the Royal United Services Institute on 15 March 1985, warned of the danger of delaying consideration of the strategic implications of the SDI research programme:

Can we afford even now simply to wait for the scientists and military experts to deliver their results at some later stage? . . . I do not believe so . . . research into new weapons and study of their strategic implications must go hand in hand. Otherwise, research may acquire an unstoppable momentum of its own, even though the case for stopping may strengthen with the passage of years. Prevention may be better than later attempts at a cure. We must take care that political decisions are not pre-empted by the march of technology.

This is a very difficult subject on which to offer proof or refutation. Without the scientists, the arms race would certainly come to a stop, but, as the quotation from the Foreign Secretary's speech suggests, that does not imply that they alone are the prime movers in it. As we have indicated, the administrative machinery for weapons procurement in the UK is certainly not geared in that way, unless by 'the scientists' is meant everybody in the whole military R&D complex – in the MoD, the Armed Forces, and industry. Some projects, such as Chevaline perhaps, do seem to be pursued far more persistently than their practical objectives would appear to justify. There has, however, never been any doubt that the final, formal decision about them rests with the politicians rather than with the civil servants, engineers, or scientists who are directly involved, although if faced with a united front of technical and military advice, it would be a brave Secretary of State who could resist it.

In any case, it is hardly fair to blame people individually for showing great zeal and ingenuity in the invention or improvement of weapons, when this is precisely what they are employed to do. However humanitarian their personal impulses, they are caught in the logic of their situation. Once one has actually thought of a peculiarly effective weapon, however horrible, one has to reckon with the possibility that one's opposite number in an R&D establishment of a potential enemy may also have thought of it and is secretly developing it for use. This acute moral dilemma was aroused during the Second World War, in relation to the atomic bomb, and has been characteristic of military R&D ever since.

On the other hand, the scientists in each country all belong, despite themselves, to an international system of *competing* military R&D organizations, governmental and commercial, whose innovative power puts an immense burden on the people and nations of the world. Political realism might insist that this system is fuelled by the merciless tensions and rivalries of those very nations, and is not an independent institution. Nevertheless, the larger the size of this system, the more people who are involved in it, and the more money that is spent on it, then the faster the rate of innovation and the heavier its burdens become. The world would benefit as a whole if military R&D were substantially reduced across the board. This cannot be done by unilateral action, and no nation of the size, geography, and history of Britain can afford to withdraw altogether from this competition, but it can take care to keep its effort as small as is consistent with clearly defined military and political requirements.

One of the effects of the general moral sentiment against military R&D is to isolate its practitioners from other scientists and engineers, especially in academic work. As we have noted, this is also a consequence of the way in which it is organized and funded in the UK; very little university research is directly supported by the MoD (even the MoD initiative, announced in the 1985 *Statement on the defence estimates*, to engage in a co-operative research scheme involving the research councils and the universities, amounts to only £10 million per year), and very few university scientists are involved as consultants in military projects. Many people (including the Association of University Teachers) support this separation on broad ethical grounds. They also point to the problems of maintaining technical secrecy, and other forms of national security, in institutions whose traditions are all of scholarly openness and cosmopolitanism. Whatever their own personal stand on such matters as nuclear disarmament or the Strategic Defense Initiative, they would prefer to keep these tense political issues out of academia.

But there is another side to this argument. The defence of the realm is a vital aspect of national life, and its problems and necessities should not be shunted out of the way and left to a few specialists. A moralist might argue that there is more virtue in getting involved in a dubious activity and trying to reform it from the inside than standing aloof and denouncing it at no cost to oneself. And from a purely pragmatic point of view, people such as university scientists who participate occasionally in military R&D are in a much better position to provide independent advice and criticism of policies and practices than full-time practitioners who depend on the organization for their livelihood. This is the situation in the United States, where the public debate on weapons procurement and military strategy is far better informed than it is in Britain. We should consider very carefully whether our practices are superior to theirs in this respect.

5 Conclusions and recommendations

Our study has not uncovered grave deficiencies in military R&D in Britain, calling for urgent, radical action. But we have come to the conclusion that present policies and practices should be revised over the next few years in four significant aspects.

1. Reduction of overall effort and concentration of resources.

2. Closer relation with other sectors of national scientific and technological activity.

3. Greater attention to policies to curb the arms race.

4. More public accountability.

5.1 SIZE AND EFFICIENCY OF MILITARY R&D EFFORT

In our view, Britain's military R&D, although geared to military roles determined by government, is excessive in relation to her economic status (see Chapter 2). This is due to Britain's position up to, and indeed for a while after, the Second World War, as one of the world's leading nations in the design and development of most weapons and other military equipment. That position in turn arose from the extensive set of military commitments, and hence equipment requirements, which Britain then had.

As the military commitments were reduced in scope after the Second World War through the loss of the Empire and, later, the abandonment of the East of Suez role, so Britain's military requirements became increasingly similar to those of the other medium powers in Europe. Yet the governmental and industrial infrastructure for equipment development and production shrank more slowly; there were always new threats to which to react, and ageing equipment to replace, and somehow the range of Britain's military R&D and production remained unduly large.

Financial pressure is now bearing down hard on the overall defence programme and, within it, on the R&D element. We believe that, although any government will naturally strive for as much self-sufficiency in the military equipment field as it can afford, the time has come for a clearer and more explicit view to be taken of the cost and benefits of, and the balance to be struck between, the following four options.

1. The maximum possible degree of self-sufficiency (Paras. 4.3.1 and 4.2.2).

2. International collaboration (arranged to maximize efficiency rather than political equality) (Para. 4.3.3).

3. Buying from abroad, manufacture under licence, or co-production (Para. 4.3.1).

4. Abandonment of a major military role (Para. 4.2).

A continuing high degree of self-sufficiency will only be possible if much greater productivity and efficiency can be extracted from the current R&D and production system. But while there is always room for improvement, in particular through use of competitively determined fixed price contracts (Paras. 3.2.2 and 4.3.2), it is hard to see where gains of the necessary magnitude might be obtained.

1. The defence research establishments (Para. 3.2.1) have now been more or less reduced in numbers, staff, and functions to the tasks which the Strathcona Committee[1] defined as those which fall most naturally to government, and from which government could not normally divest itself. These were: 'concept formulation; generation and vetting of specifications; acceptance and evaluation; and certain statutory tasks like aircraft airworthiness'.

2. Just about all the tasks which Strathcona believed could in principle be done in industry or other non-government bodies – namely: design, development, project support, post-design services, and certain research tasks – have now been, or are being, transferred to industry, including the Royal Ordnance Factories.

3. While policies to stimulate industrial efficiency, such as the current emphasis on competitiveness (Para. 4.3.2), should be encouraged within reason (that is, they must not be pushed to the point of driving key firms out of the defence business unless as a deliberate act of policy), there is no evidence that they can yield *massive* savings compared with well-run, non-competitively awarded contracts. Nor is it obvious that the transfer to industry of many of the functions previously performed in the defence research establishments will result in better design or more export-oriented designs: there is no evidence that industry is any better than MoD staff in predicting what overseas customers, who frequently buy 'off the shelf', will want in 10 years time.

Accordingly, if Britain is to continue to produce military equipment which could survive in a wartime environment, the character of which would be determined by the capabilities of the (much richer) superpowers, it must concentrate its R&D resources. We must emphasize that in saying this we are not advocating a reduction in commitment to defence *per se*. Paradoxically, in fact, a reduced and more concentrated R&D effort might improve Britain's defensive posture by releasing funds for equipment purchase. It is in this context, then, that we see the need for either greater collaboration (so as to share R&D costs), more dependence on imports, manufacture under licence or co-production (so as to buy in R&D, in effect), or the abandonment of a major military role and its attendant equipment needs. Because we consider the last of these to be unlikely in the immediate future (and even the abandonment of the nuclear role would involve transitional R&D – see Para. 4.2), the key choice is between collaboration, on the one hand, and imports, manufacture under licence, or co-production on the other.

Collaboration (see Para. 4.3.3) could undoubtedly be increased. It has been suggested[2] that all projects costing more than, say, £250 million should be automatically proposed as collaborative projects, and that Britain should seek agreement with its main industrial allies that an, at first, limited number of large projects should be put out to competition between international consortia. Mr Heseltine has in fact instructed MoD staff to refer all projects of significance to ministers at the Staff Target stage so that the international collaborative possibilities may be considered from the outset. A discussion of collaborative possibilities has now become an integral part of the progress of a project within MoD. At the same time, we would argue, the commitment phase of any collaborative proposal should be handled with very great circumspection, for not only does its outcome determine the management arrangements for the project, on which so much will later turn, but also, once the commitment has been entered into, escaping from a collaborative project is normally far harder than cancelling a purely national one. Likewise, collaborative commitments for R&D entered into at the level of firms (and not government) could, in time, generate political pressure upon government either to proceed with production, or at least not to be critical in public of the R&D programme. This is a possibility that must be watched with reference to the involvement of British firms in the Strategic Defense Initiative.

Imports, manufacture under licence, and co-production, would involve the rundown of some R&D and design, if not also production, facilities. Many of the people currently employed in those facilities have acquired scarce and valuable skills (Para. 4.5). These skills are often difficult to apply *immediately* in other types of R&D; nevertheless, given time, material resources, and sympathetic management, almost all the highly specialized scientists and engineers in government and industrial military R&D establishments are capable of earning far more than their keep in the civil economy and contributing markedly to the wealth of the nation. Should it ever be decided, for example, that the need for the Chemical Defence Establishment (Porton Down) had ended, CDE could be converted to a research association for the protective clothing and environmental monitoring industries. To ensure that this valuable national resource is not wasted it would be essential for an organized programme of diversification (or conversion) and redeployment to be introduced. Such a programme would need to be financed over a considerable period of time – upwards of a decade. It could draw upon previous British experience of redeployment in general (such as at the end of the Second World War) and in government research establishments (notably Harwell and the Microbiological Research Establishment) in particular, as well as the detailed plans for industrial conversion produced by such bodies as the Lucas Aerospace workers and the Vickers shop stewards, and the work that has been done on this subject by the Institution of Professional Civil Servants (IPCS).[3] Much of the pressure for maintaining the current range of R&D and production facilities comes from the people employed in those facilities, who naturally fear for their own future; progress in reducing the scope of Britain's military equipment development and production can be maximized only if it has the consent not only of the taxpayer but also of the workers and unions in the relevant industries and, within the public sector, of the IPCS. We are confident that this could be obtained provided that government formulated a coherent long-term conversion strategy.

5.2   RELATION TO NATIONAL SCIENCE AND TECHNOLOGY POLICY

Elementary economic and political rationality suggests that funds for military R&D should be allocated only after some consideration of their impact on the total national effort in science and technology. In our opinion, the very large share (Para. 2.3) of the military sector in overall R&D activity makes this reform imperative. But the very notion of a national science and technology policy is strongly resisted in Whitehall and Westminster.[4]

British government is organized on the basis of the responsibility of ministers for their departments. Therefore, any cross-cutting measures designed to co-ordinate along any one dimension of departmental activity, such as R&D, would be said to confuse the lines of accountability. Moreover, British government does not think of its R&D spending as a single cake waiting to be cut and distributed to departments. The orthodox view is that R&D is budgeted for in separate slices, one from each departmental cake. These cakes are so diversely composed that an aggregate of all the R&D slices would have little administrative or financial significance; the items would be so distinct and different that none of them could be converted into another. Hence, the argument runs, the volume of military R&D spending is set to meet the needs of the Ministry of Defence. It does not draw funds away from other R&D activities; it competes only with other *defence*-related activities, and any reduction in military R&D would have no automatic consequences for the rest of governmentally-funded R&D.

But even if the R&D slices of the baked cakes are not interchangeable, to some extent their ingredients prior to cooking are. More plainly, even if R&D funds are a somewhat elastic quantity, R&D personnel are less so, at least in the short-term. Only a finite stock of skilled people exists, coupled with a flow of new graduates that can be increased or decreased only slowly. Failure to manage the division of these people between the civil and military sectors according to whatever constitutes the national priorities of the day could be damaging.

We note that the House of Lords Select Committee on Science and Technology, the Advisory Council for Applied Research and Development (ACARD), and the Advisory Board for the Research Councils (ABRC) (in particular through the Mason report) have all in recent years urged better central co-ordination of one or another aspect of science and technology policy.[5] So also has the Council for Science and Society in an earlier report,[6] while more recently the Trades Union Congress has called for a national strategy for technological innovation, to include a switch of resources from the military to the civil sector.[7] The Prime Minister's Lancaster House seminar of September 1983, and the Cabinet Office's Annual Reviews of R&D, suggest that this policy path is not entirely blocked in high places.

We are not necessarily advocating the establishment of new administrative structures. More could be done without any changes in machinery. The ABRC could press ahead with implementing the Mason report's recommendations that the Board should have before it the research plans of all government departments (including Defence) when it considers its recommendations concerning the research councils. ACARD, similarly, could take a more persistent interest in military R&D. We strongly commend such developments.

5.3 THE ARMS RACE AND ARMS CONTROL

Military R&D is closely connected with the problems of the arms race and arms control. One view of the arms race is that it is driven by new technology, either in the form of an 'action–reaction' spiral, or by a momentum that derives from the inherent dynamism of technology itself, regardless of what potential adversaries are doing. Neither variant, on analysis, offers an adequate explanation of the arms race. Nor do wholly domestically-focused explanations, such as the military–industrial complex thesis, or wholly internationally-focused ones, such as those which explain the arms race in terms of instabilities in the international system. Any general explanation must involve a mixture of all these factors, and more.

Yet military R&D, and the forces which influence its shape and size, remain major contributors to the arms race. (cf. Para. 4.7). This is true even though not all military R&D necessarily so contributes. Technologies which increase stability in a crisis, and deter the adversary while reassuring one's own 'side', must surely be distinguished from those which (as many have said of the Pershing-2 missile, for example) increase crisis instability, scare their own side, and invite attack.

British military R&D, being smaller in scale and scope than that of the superpowers, clearly has much less effect on the arms race than theirs do. In general, Britain is more of a follower than a leader in this field, especially in large-scale and radical innovations, such as 'Star Wars'. Her international role as a major developer and exporter of advanced weapons remains considerable, however, so that even if British equipment follows the lead of the superpowers (as it has to if it is intended that it survive the first superpower attack upon it), its role in the diffusion of weapons to other parts of the world remains substantial. On that point, as we have already recommended, decisions to export should be based on overall considerations of national policy, which is not necessarily identical to simple opportunism in the export trade (Para. 4.6).

More positively, Britain could play an increasing role in international thinking about the implications for arms control of certain current trends in military R&D. One such trend is the convergence between civil and military developments in the microelectronics field, a convergence which official policy is encouraging. A second is the increasing emphasis upon design of weapons and platforms for multirole use (nuclear capable artillery; dual nuclear/conventional arming of cruise or Pershing missiles). Even without these tendencies, it would have been awesomely difficult to come up with arms control measures that might curtail the development of new weapons, though a United Nations expert group on military R&D has been grappling with just this problem.

There is, however, scope for a country with as extensive a military R&D programme as Britain's to take a lead in developing thinking about such problems. Examples of what is needed include investigations of verification possibilities for dual-capable weapons, social science research into problems of arms control (such as national perceptions of each other, reasons for particular weapons build-ups, and negotiating strategies), and development of systems for non-offensive defence. There would also be value, from the point of view of informed public debate, especially in the context of the 'nuclear winter' question, of adding the Ministry of Defence's own undoubted expertise on the effects of nuclear weapons to that debate; the Home Office's handling of technical questions on this subject has so far been lamentable. These are all questions which the Arms Control Unit in MoD should take up.

The Unit should also watch carefully to see that the British government's sceptical position on the US Strategic Defense Initiative does not become hamstrung by substantial industrial dependence upon that initiative. Great divisions exist within NATO on the strategic significance of the SDI, and upon the possible implications of its claims upon resources for more immediate R&D and equipment programmes. It would be tragic if the British government allowed understandable industrial desire for a share of the SDI R&D funds to cloud its judgement about the Initiative. In our opinion, the SDI is, in the long run, a strategic blind alley. If we could be certain that its success rate would be 100 per cent, then we might be inclined to back it, but as it is, and since in the arena of defence against nuclear weapons success rates of less than 100 per cent will be unacceptable, we remain sceptical. The question of the boost to technology, especially in the electronics and computing fields, that is likely to result from the SDI, must be kept separate from the strategic implications of the Initiative. If the British or other European governments want their own industries to receive a similar boost, they must seek ways of funding them that do not entail entering this blind alley (Para. 4.4.3).

5.4 PUBLIC ACCOUNTABILITY AND SECRECY

Military R&D is conducted largely out of sight of the public and their elected representatives. Security considerations naturally make this sector of public policy somewhat different from others. Even so, given the costs and other implications involved, it is reasonable to ask that ministers and MPs should have good access to the details of defence equipment programmes, and that they should have this access *before* programmes hit the headlines on grounds of cost over-runs, as Chevaline and the Nimrod Airborne Early Warning aircraft, to take but two prominent examples, have done in recent years.

Very little information is available on ministerial involvement with military R&D. One viewpoint is that of Dr David Owen, who warned that

The insidious processs of military indoctrination, a heady mixture of pomp and secrecy to which most politicians involved in defence are susceptible, tends to blunt one's normal sensitivity. One can easily become part of the very military machine that one is supposed to control.[8]

In the case of the nuclear programme, there is some evidence that ministers have always been in control of new developments, although less so for Chevaline than for the other cases.[9] Chevaline also shows the capacity of ministers to exclude projects from the normal channels of debate and discussion, even within Whitehall. Successive secretaries of state have expressed concern about the volume and range of R&D spending, and have instituted reviews of both procedures and programmes. Whether, however, ministers have been sufficiently involved in regular and systematic review of the R&D programme, especially in the early stages of projects, is doubtful.

In a formal sense, parliament also has opportunities, if not to control military R&D, then at least to exact a measure of accountability over it. Parliament is, however, notorious for its limited capacity to deal with technical subjects. The same is true for long-term projects. Both ministers and MPs, but especially MPs, naturally tend to focus on events with time-scales within the lifetime of the present parliament, a tendency which weakens the rendering of account for all advanced technological projects (cf. Concorde) and not only military ones. Nevertheless, through its select committees, Parliament potentially has the means of gathering detailed information on, and making thorough reviews of, military R&D projects. Both the Public Accounts Committee and the Defence Committee have used this capacity to advantage on numerous occasions, and in so doing have added valuably to public debate on these matters.

There are, however, substantial weaknesses in parliamentary scrutiny. The Public Accounts Committee can only act retrospectively. It does not have access to sufficiently detailed accounts to be able to discover the existence of such projects as Chevaline until the government chooses so to inform parliament. Nor can it compel the government to change any of its practices; it can exert such influence as it has only through embarrassment, or threat of embarrassment, of ministers and permanent secretaries. The Defence Committee can deal with current and proposed aspects of defence policy, but again, as with all parliamentary select committees, it has no power to block executive proposals, nor even to obtain all the information that it desires. The first point was demonstrated when, during Defence Committee hearings into Trident in 1980–81, the Government invited the House of Commons to endorse the Trident decision without awaiting the Committee's report. As for the second point, the Defence Committee has expressed dissatisfaction with the quality of information provided to it by MoD during hearings in 1984–85 on defence equipment. In the Committee's words,

We wanted to examine in-service dates, operational life, scale and phasing of expenditure. Our purpose was frustrated by vague and evasive answers and elegant but unhelpful hypotheses. Our experience in previous inquiries has not led us to expect the Ministry to volunteer information on matters which may be politically charged or potentially embarrassing; but we do expect proper answers to questions which are asked as part of our task of examining 'the administration, policy and expenditure' of the Ministry of Defence.[10]

We accept that some level of secrecy over R&D on military projects is necessary. However, too much secrecy is detrimental to efficiency, and lack of public knowledge of even the existence of a project makes it more, rather than less, difficult for responsible politicians to blow the whistle when necessary. Chevaline is a sad example of this. This project was continued for 14 years at a total cost of over £1 billion before the House of Commons was informed of it. Its original estimated cost in 1972 was £175 million. For over a decade, under the strictest secrecy, the project limped along behind schedule and increasingly over budget, a catalogue of bad management, lack of control, and indecision. Large sums of money which could have proved vital for other sectors of the defence budget were frankly wasted.

We should suggest various changes to give better parliamentary scrutiny and control over military R&D expenditure.

1. All individual R&D projects requiring expenditure over £5 million in any financial year should be detailed in the annual Defence Estimates, with a statement of total estimated R&D costs and time-scale.

2. The criticism of the Public Accounts Committee in its report concerning Chevaline 'that the costs were not disclosed, and there was no requirement that they should be disclosed . . .', should now be acted upon. The powers of the Defence Select Committee and the Public Accounts Committee should be increased to give them the right to scrutinize projects in the very early stages. They must receive all information necessary to enable them to do this. (Procedures exist for the protection of commercial confidentiality and other matters requiring discretion: the Committees can, for example, and do, meet in private session on occasion.)

3. The proceedings of the Committees should in general be publicized more effectively.

4. To assist with the first three recommendations, the Ministry of Defence should publish annually a report on research and development. This should list the main projects under development, give details of costs, expected completion dates and progress, and explain the reasons for the decision to fund each of these projects. It should also give details of the main research programmes of the research establishments. Comparisons with the United States are always dangerous, but when we see the detailed documentation on R&D that is presented annually by the Pentagon to Congress, we cannot but reflect on how such a regular influx of information to Parliament could raise the level of discussion in Britain. While the detail given in the annual *Statement on the defence estimates* has improved remarkably in recent years, this is in a sense only to demonstrate how extraordinarily uninformative these statements used to be. We do not consider that the implementation of this recommendation would jeopardize security. Not only does the vast amount of published information on US military R&D give the Soviet Union a good sense of the state of the art in the West, but outline information is in any case normally available (for example, in the trade press) on development contracts in the UK. The difficulty is that this information is not conveniently accessible in one place. We are *not* suggesting that precise details of the intended characteristics of weapons systems be published.

The final requirement for increased public accountability, based on a more informed public and Parliament, goes wider than these recommendations. What again is lacking in Britain, compared with the United States, is an extensive community of independent analysts of military R&D policy. The subject has not been much studied in the universities or the independent research institutes. We have come to the conclusion that policy-making would be significantly improved in the long term if the traditional isolation of military R&D from other sectors of academic and scientific life were deliberately reduced. There is no tradition in Britain, unlike the USA, of rotating scientists and social scientists into government and then back out. Hence, Britain lacks the critical mass of informed expert commentators which would make an intelligent debate possible. Yet, it will not do to rely on the experts *within* MoD, good though the Ministry's internal systems for testing arguments are. Not all officials could be expected to agree with us, though the more farseeing ones may. Ministers, however, and politicians generally, of all parties, should surely see it as in their interest to encourage the development of an expert community of independent military R&D analysts who could help raise the standards of public debate on this crucial subject. Defence generally, and defence R&D in particular, use scarce resources; they require more extensive and informed analysis than they have enjoyed in the past.

References

CHAPTER 2

1. See Stockholm International Peace Research Institute (1976). *Resources devoted to military research and development.* In *World armaments and disarmament: SIPRI Yearbook 1972*, pp. 155–64. Almquist and Wiksell, Stockholm.
2. Cabinet Office (1984). *Annual review of government funded R&D 1984*, p. 41, Table 7.1. HMSO, London.
3. The scheme was announced in Ministry of Defence (1985). *Statement on the defence estimates 1985*, paragraph 537. HMSO London. The figure of £10 million was given in *The Times Higher Education Supplement*, 12 July 1985.
4. Organisation for Economic Co-operation and Development (1984). *OECD science and technology indicators: resources devoted to R&D*, pp. 118 and 128. OECD, Paris.

CHAPTER 3

1. *The Economist*, 24 March 1984, p. 20.
2. Turpin, C. (1972). *Government contracts*, Chapter 6. Penguin Books, Harmondsworth.
3. House of Commons (1985). Note by the Ministry of Defence, in House of Commons, Third Report from the Defence Committee session 1984–85, *Defence commitments and resources and the defence estimates 1985–86, Volume III: the defence estimates 1985–86*, HC 37-III, pp. 26–7. HMSO, London.

CHAPTER 4

1. Gowing, M. (1974). *Independence and deterrence: Britain and atomic energy 1945–1952*, Vol. I, *passim*, but especially p. 184. Macmillan, London.
2. But see Simpson, J. (1983). *The independent nuclear state: the United States, Britain and the military atom*, Chapter 10. Macmillan, London.
3. Chapter 4, ref. 2, pp. 157 and 171.
4. Institution of Professional Civil Servants (1984). *Nuclear arms, defence spending and jobs*, paragraphs 39 and 40. IPCS Document NEC/GEN/75/84, London.
5. Kaldor, M. (1982). *The baroque arsenal*. Andre Deutsch, London.

6. Ministry of Defence (1982). *Statement on the defence estimates 1982*, Cmnd. 8529-I, paragraph 411. HMSO, London.
7. Chapter 4, ref. 6, paragraph 430.
8. *The Economist*, 25 February 1984, pp. 30–1.
9. Hartley, K. (1984). Competition and UK defence policy. *Economic Affairs*, January 1984, 16–18.
10. Ministry of Defence (1984). *Statement on the defence estimates 1984*, Cmnd. 9227-I, paragraph 237. HMSO, London.
11. Edmonds, M. (1977). Defence as a declining industry: the British experience 1964–77. University of Lancaster, unpublished paper presented at Colloque sur Tradition et Changement dans les Systèmes Militaires Occidentaux, Sorèze, 25–27 July 1977.
12. Callaghan, T. A. (1975). *US–European economic co-operation in military and civil technology*. Centre for Strategic and International Studies, Georgetown University, Washington DC. (revised.) Cited in Hartley, K. (1983). *NATO arms co-operation: a study in economics and politics*, p. 4. Allen and Unwin, London.
13. van Houwelingen, F. (1984). The Independent European Programme Group (IEPG): the way ahead. *NATO Review*, **32** (4), 17–18.
14. Callaghan, T. A. (1984). The structural disarmament of NATO. *NATO Review*, **32** (3), 21–6.
15. This list is adapted from Deitchman, S. L. (1979). *New technology and military power: general purpose military forces for the 1980s and beyond*, pp. 197–8. Westview Press, Boulder, Colorado.
16. Three good sources on this subject are Hartley, K. (1983) (ref. 12 above); Edmonds, M. (Ed.), (1981). *International arms procurement: new directions*. Pergamon, Oxford; and Williams, R. (1973). *European technology (the politics of collaboration)*. Croom Helm, London.
17. Hartley, ref. 12 above (Chapter 4), p. 161.
18. Chapter 4, ref. 6, paragraph 407.
19. Ministry of Defence (1981). *Statement on the defence estimates 1981*, Cmnd. 8212-I, p. 48. HMSO. London.
20. Taylor, T. (1980). Research note: British arms exports and R&D costs. *Survival* **22** (6), 259–62.
21. House of Commons (1982). Second report of Defence Committee session 1981–82. *Ministry of Defence organisation and procurement*, HC 22, Vol. I, paragraphs 74–75; Vol. II, pp. 306–7, and question 1600. HMSO, London.
22. Electronics Economic Development Council (1983). *Civil exploitation of defence technology*, Report to the Electronics EDC by Sir Ieuan Maddock, and Observations by the Ministry of Defence. National Economic Development Office, London.
23. House of Lords, Select Committee on Science and Technology, 1982–83, (1983). *Engineering research and development*, HL 89, paragraphs 19.1 and 19.2. HMSO, London.

24. DeGrasse, R. W. Jr, (1983). *Military expansion economic decline*, pp. 102–5. Council on Economic Priorities, New York.
25. Chapter 4, ref. 24, p. 105.
26. Chapter 4, ref. 24, pp. 106–18.
27. Noble, D. F. (1982). The social and economic consequences of the military influence on the development of industrial technologies. In *The political economy of arms reduction: reversing economic delay*, (ed. L. J. Dumas). Westview Press, Boulder, Colorado. Reppy, J. (1982). Influence of military R&D on civilian technology and trade. Cornell University. Unpublished paper presented at the Midwest Political Science Association, Milwaukee, Wisconsin, April 1982.
28. Lichtenberg, F. R. (1984). The relationship between Federal contract R&D and company R&D. *American Economic Review* **74** (2), 73–8.
29. Soete, L. and Dosi, G. (1983). *Technology and employment in the electronics industry*, Chapters 2 and 8. Frances Pinter, London.
30. Taylor, J. M. (1984). Computing and software. In The Economist Intelligence Unit, *Defence papers: a transatlantic debate over emerging technologies and defence capabilities*, pp. 60–1. The Economist Intelligence Unit, London.
31. *Review of the Scientific Civil Service (1980)*, The Holdgate Report, Cmnd. 8032, 1980, paragraphs 4.5 and 4.6, and Table 3. HMSO, London.

CHAPTER 5

1. Ministry of Defence (1980). *Steering Group on Research and Development Establishments*, The Strathcona report, paragraph 10. Ministry of Defence, London.
2. *The Economist*, 25 February 1984, pp. 30–1.
3. For discussion of conversion possibilities, see Dumas, L. J. (Ed.), (1982). *The political economy of arms reduction: reversing economic decay*. Westview Press, Boulder, Colorado. This book contains details of the Lucas case; and Kaldor, M., Smith, D. and Vines, S. (Eds.), (1979). *Democratic socialism and the cost of defence*. Croom Helm, London for discussion of Lucas, Vickers, and other cases.
4. See Gummett, P. (1980). *Scientists in Whitehall*, Chapter 7, Manchester University Press, Manchester.
5. For example, House of Lords, (1981). First report of select committee on science and technology, Session 1981–82. *Science and government*, HL 20-I. HMSO, London; Advisory Board for the Research Councils (1983). *A study of commissioned research*, The Mason Report. ABRC, London.
6. Council for Science and Society, (1982). *Technology and government*. CSS, London.
7. Trades Union Congress (1985). *The future business: Britain's research and development crisis*. TUC, London.

8. Owen, D. (1972). *The politics of defence*, p. 14. Jonathan Cape, London.
9. See Chapter 4, ref. 2, pp. 232–4.
10. See Chapter 3, ref. 3, Vol. I, paragraph 36.
11. House of Commons (1982). Ninth report of the Committee of Public Accounts Session 1981–82. *Ministry of Defence: Chevaline improvement to the Polaris Missile System*, HC 269, paragraph 15. HMSO, London.